Agricultural Growth, Farm Size and Rural Poverty Alleviation in India

Professor Chennamaneni Hanumantha Rao is currently Chairman, Board of Governors, Institute of Economic Growth, University of Delhi and Centre for Economic and Social Studies (CESS), Hyderabad. He is also Member, National Advisory Council, Government of India.

Born in 1929 in Karimnagar, Andhra Pradesh, he did his graduation from Nizam College, Hyderabad. As an activist of the All Hyderabad Students Union, he participated actively in the freedom struggle. He did his M.A. in Economics from Osmania University, Hyderabad (1957) and his Ph.D. from University of Delhi (1962). He was a post-doctoral fellow at the University of Chicago during 1966-67. He was a Fellow at the Institute of Economic Growth, University of Delhi (1961-92) and also its Director (1976-80).

In addition to a distinguished academic career, Professor Rao has been at the forefront of several high-level policy making bodies under Government of India. He was Member, Planning Commission (1982-86); Member, Seventh and Eighth Finance Commissions; Member, Economic Advisory Council to the Prime Minister (1983-85); Chairman, National Commission on Rural Labour (1990-91); Chairman, Technical Advisory Committee on DPAP & DDP, Ministry of Rural Development (1993-94); Member, Technical Advisory Committee (TAC) of the Consultative Group on International Agricultural Research (1995-97); Chairman, High Powered Fertilisers Pricing Policy Review Committe (1997-98); and Member, Board of Directors, Reserve Bank of India (1994-2000).

His publications include "Technological Change and Distribution of Gains in Indian Agriculture" (Macmillan, 1975); "Agricultural Growth, Rural Poverty and Environmental Degradation in India" (OUP, 1994); "Andhra Pradesh Development: Economic Reforms and Challenges Ahead" (edited with S. Mahendra Dev, CESS, 2003) and over 100 research papers in academic journals.

Professor Rao has received many awards during his illustrious career. These awards/honours include Rafi Ahmed Kidwai Memorial prize (1974-75), Sri Kambhampati Satyanarayana (Senior) Memorial Award (1998), K.H. Batheja Award (1994), Financial Express Award (1995), Honorary Degrees of Doctor of Philosophy from Srikrishna Devaraya University, Anathapur (1991) and the Kakatiya University, Warangal (1998), in Andhra Pradesh; and Sri Ravi Narayan Reddy Memorial Award (1994), and Telgu Atma Gaurava Puraskaram (2000). He has been awarded 'Padma Bhushan' by the President of India in 2004.

Agricultural Growth, Farm Size and Rural Poverty Alleviation in India

Selected Papers by

C H HANUMANTHA RAO

ACADEMIC FOUNDATION
NEW DELHI

First published in 2005
by
ACADEMIC FOUNDATION
4772-73 / 23 Bharat Ram Road, (23 Ansari Road),
Darya Ganj, New Delhi - 110 002 (India).
Phones : 23245001 / 02 / 03 / 04.
Fax : +91-11-23245005.
E-mail : academic@vsnl.com
www : academicfoundation.com

Cover photographs, courtesy :
Directorate of Extension, Ministry of Agriculture, IARI Campus, New Delhi.

Agricultural Growth, Farm Size and Rural Poverty Alleviation in India
Selected Papers by
C.H. Hanumantha Rao

ISBN 81-7188-419-9

Designed and typeset by Italics India, New Delhi.
Printed and bound in India.

To

Professor A.M. Khusro

CONTENTS

...CONTD. ...

II

Farm Size and Tenure

III

Rural Poverty Alleviation

List of Figures and Tables

Figures

Tables

...CONTD. ...

Introduction

This book brings together a number of my papers on Agricultural Growth, Farm Size and Tenure, and Rural Poverty Alleviation in India. They were written over a period of four decades while I worked at the Institute of Economic Growth, Delhi, between 1961-1992, and thereafter, as an Honorary Professor, at the Centre for Economic and Social Studies, Hyderabad. The following discussion highlights the major issues dealt with in these papers, the relevant findings and the evolution of policies for agricultural development and rural poverty alleviation in India over this period.

Growth

At the time of independence, India inherited a fragile agriculture with declining trend in foodgrains output, low development of infrastructure like irrigation, outmoded land tenure systems and hardly any institutional framework for the supply of credit and inputs to the farmers. The First Five Year Plan, therefore, aimed at revitalising agriculture through sizeable investments in irrigation, reform of land tenures, development of institutions for the supply of credit and inputs to the farmers and launching of Community Development Programme to raise farm productivity and employment.

As a result, during the first decade of planning, there was a large expansion of area under cultivation with a significant increase in area under irrigation, and the growth of foodgrains output outpaced the accelerated population growth. However, the Second Five Year Plan instead of giving adequate priority to investments in irrigation and the development of technology, relied excessively on co-operative farming and community development for raising agricultural output - especially marketed surpluses of foodgrains. In this period , there was an interesting debate between the 'structuralist' school, i.e., those believing in drastic changes in land relations as a *pre-requisite* for effecting a breakthrough in agriculture and those who thought, based on the evidence of high output response to irrigation and new technology, that high agricultural growth was possible through input intensification, despite the smallness of farm size and the prevalence of tenancy.

It became clear by the mid-1960s that the potential for agricultural growth from the investments made in irrigation, and from the autonomous factors, e.g., the expansion of area under cultivation, incentives created by the implementation of land reforms like the abolition of intermediaries, and the rise of agricultural classes to political power were nearly exhausted, and that further growth of agriculture depended crucially on the expansion of agricultural infrastructure and the application of new technology to raise farm productivity and profitability. The large imports of foodgrains in the wake of two successive droughts in the mid-sixties, and the unacceptably high political costs that it entailed, pushed the government towards a bold strategy for achieving self-sufficiency in foodgrains through the green revolution by stepping up investments in irrigation, evolving and applying high-yielding varieties of seeds and intensifying the use of inputs like fertilisers.

This strategy paid high dividends with the country achieving near self-sufficiency in the availability of foodgrains within a period of less than two decades. Beginning from the late 1960s through 1970s there was a steep rise in wheat output and by the end of 1980s there was a breakthrough in rice output in large parts of the country, notably in the eastern Gangetic Plains. For over a decade now the country has been a witness to the paradox of large 'surpluses' of foodgrains coexisting with high incidence of poverty and malnutrition. This situation, in retrospect, underscores the continued relevance of some of the concerns voiced by the votaries of the 'structuralist' school regarding the adverse socio-economic consequences of green revolution focused exclusively on raising 'productivity' in the infrastructurally developed regions to the neglect of greater employment generation and purchasing power for the poor, especially in the less developed regions.

The scope for extending the known technology for raising foodgrains output has by no means been exhausted. There is still a large gap between the attainable potential with known technology and the actual output in several parts of the country, particularly in the eastern region. This potential can be tapped by the provision of complementary inputs like irrigation, strengthening extension services and the supply of institutional credit. Rain-fed or 'unfavourable' areas too offer a significant scope for increase in farm output through the development of infrastructure and the use of emerging biotechnologies like genetic engineering of seeds, tissue culture, etc. Indeed, investments in them for unfavourable areas are strongly indicated by the high profitability of such investments and greater scope for poverty alleviation. On the other hand, in regions like North-West where green revolution has been a success and limits to growth in cereal output have been reached from the available technology, policy has to focus on greater diversification of agriculture, agro-processing, and the growth of rural non-farm enterprises.

Stability

Despite an increase in the proportion of area irrigated, there is no evidence, as yet, of a decline in the weather-induced fluctuations in crop output. This is attributable, in part, to the rise in the sensitivity of output to soil moisture with the application of new technology and its spread under rainfed conditions. Also, droughts have become increasingly severe on account of over-exploitation of ground water and degradation of land leading to the shortage of drinking water, fodder and fuel. Thus, whereas the government has been able to accumulate adequate stocks of foodgrains to support the public distribution system, the rural poor in dryland areas continue to experience hardships on account of drastic reduction in employment and purchasing power in the event of failure of rains.

Crop insurance, against yield risks, introduced in the country has not been a viable proposition and had to be heavily subsidised because of widespread damages to crops whenever there is a failure of rains. Since the premia payable by the farmers under such circumstances have to be quite high, investments for avoiding risk such as in irrigation may be preferable to the heavily subsidised crop insurance. Recent studies indicate that rainfall insurance could be superior to crop yield insurance, particularly because it is much simpler and provides safeguards against moral hazards associated with crop insurance. However, even this may not be attractive to private insurers unless the bonds issued against rainfall events could be appealing to international investment bankers.

At the same time, it would be desirable to emphasise, in addition to the development of irrigation, strategies that mitigate the adverse consequences of drought such as regionally and seasonally broad-based growth of agriculture with greater diversification and due importance given to small farmers who are able to realise higher cropping intensity by using their abundant family labour. Such a strategy would contribute to smoothening the fluctuations in the annual flow of aggregate farm output on account of compensatory movements in output as between different farm enterprises, seasons and regions.

Further, experience has shown that the expansion of the on-going programmes for soil and water conservation on a watershed basis through the development of grass-roots institutions for the effective participation of people would go a long way towards drought-proofing.

Price Policy

In the pre-green revolution period, owing to the chronic shortage of foodgrains and the rising prices, the price policy for foodgrains became consumer oriented. The concern for protecting the consumers predominated over the need for ensuring incentive prices for producers, partly because the ruling prices for producers tended to be high. The Single State Zonal System for foodgrains clamped by quite a few states - mostly in the pre-green revolution period - sought to protect the consumers of the respective states without due regard for the welfare of consumers in the neighbouring states or for the incentives to producers within such Single State Zones.

However, with the progress of green revolution, the need to ensure minimum support prices to producers, particularly in respect of wheat and rice, became a major policy concern. Such a policy took a definite shape towards the late sixties and the early seventies, which provided incentives to the farmers for the application of technologically new inputs and thus contributed to the success of green revolution in the country. However, over a period of time, price policy became highly producer-oriented through open-ended procurement of cereals like wheat and rice at minimum support prices, albeit, in a few prosperous regions. This practice undermined the incentives for diversification of agriculture and depressed the consumption of foodgrains by the poor. The emergence of embarrassingly large 'surpluses' of foodgrains with the government, particularly since the mid-nineties, despite the slow growth of foodgrains output and widespread incidence of poverty and malnutrition, is explained, in no small measure, by the political pressures to peg the minimum support prices persistently at levels much higher than those recommended by the Commission on Agricultural Costs and Prices.

'Surplus' Cattle

A major concern voiced by some scholars in India, particularly in the pre-green revolution period, related to the existence of 'surplus' cattle. In their view, a reduction in the number of cattle would make possible greater productivity through higher feed intake per animal. As against this, it was argued by some others on the basis of their studies that farmers appeared to be rational in the management of their cattle economy and the apparent 'surplus' was probably explained by the need for maintaining an adequate supply of drought power in a traditional agriculture subject to high risk of mortality of cattle. Also, the supply of fodder was not a severe constraint as it represented a joint product with grain on the farm itself. This 'rationality' hypothesis gained support by the experience of the post-green revolution period, especially in the eighties and the nineties, when mechanisation reduced the requirement of cattle for drought power and, consequently, the number of bullocks and of cattle in general showed a significant decline. On the other hand, the number of high-yielding milch animals and small ruminants maintained by the rural poor for supplementing their incomes increased significantly.

Post-Reform Performance and Prospects

A comparison of the performance of agriculture in India and China in the pre-reform period provides an interesting backdrop and insights for understanding their differential performance in the respective post-reform periods. Agricultural performance in terms of annual growth rate of output was surprisingly similar in both the countries for nearly three decades prior to launching of economic reforms in China in 1978. However, policy instability was conspicuous in China even as the implementation of policies chosen was far more effective than in India. The similarity of growth rates in output achieved in the two countries is deceptive in that it hides the successful implementation of land reforms in China and the failure to implement them in India. It also hides the very high rate of saving and consequently the high level of physical as well as social infrastructure development in China when compared to India. In retrospect, it is clear that the growth rate of agriculture achieved in China in the pre-reform period was much less than the potential because of factors like policy instability, adverse terms of trade imposed on agriculture and the organisation of production under communes which sapped the incentives for producers.

In the decade following the launching of economic reforms in China in 1978, when the communes were disbanded and the terms of trade were shifted significantly in favour of agriculture, the latent potential found its full expression in the achievement of a very high rate of growth of agriculture and rapid reduction in poverty. In fact, the Chinese leadership in their anxiety to provide incentives for enterprise and effort in the post-reform period, openly derided 'egalitarianism' as practiced in their pre-reform era - quite unlike in India where 'egalitarianism' has always been highly adored but rarely practiced.

In the light of the Chinese experience outlined above, the deceleration in agricultural growth as well as in rural employment growth with slow reduction in poverty in India, in the post-reform period of 1990s, is explainable in terms of unfavourable initial conditions, viz., failure to implement land reforms effectively, low rate of saving and low level of infrastructure – physical as well as social. The opening up of the economy and the significant reduction in protection to domestic industry did result in an improvement in the terms of trade for agriculture, which led to a significant rise in private investment. But because of continued decline in real public investment in irrigation, research and extension and other rural infrastructure, owing to reform-induced erosion of tax revenues and compression in public expenditure, agriculture could not derive full benefit from macro-economic reforms and globalisation, especially because, unlike in China, there was hardly any slack in the pre-reform period.

Apart from the expansion of irrigated area through enhanced public investments, efficient management of available water resources and rural electricity through the involvement of user-farmers, and regeneration of degraded lands through participatory processes, have emerged as major issues of concern in the post-reform period. Decline in public investment, inefficiency in the management of available infrastructure and degradation of natural resources are attributable, in a significant measure, to the steep rise in subsidies on water supplied from public irrigation systems, electricity for pumping water and chemical fertilisers. Public investment in agriculture now constitutes only about one-fifth of the volume of these subsidies. Political will to reduce these subsidies significantly by targeting them only to the deserving sections such as small and marginal farmers is necessary for stepping up productive investment and for conservation of natural resources.

Farm Size and Tenure

Agriculture being a biological operation and on-the-farm inputs being divisible, there is no evidence of significant economies of scale. Moreover, small farmers are better placed on account of adequate availability of family labour and closer supervision of farm operations. The experience of green revolution in the country has demonstrated that given the access to technology, inputs and credit, small farmers can fully participate in growth by adopting new technology. This experience underlines the need for public investment in agricultural research to evolve land-saving technologies and activities, adequate provision of extension services and institutional credit for small farmers to meet the growing requirements of an input-intensive agriculture. However, small farmers, and indeed farmers in general, do face problems in marketing their produce, especially in the context of diversification of agriculture. Development of marketing infrastructure and institutions such as Cooperative Marketing Societies and contractual arrangements between individual farmers and agro-processing firms can provide a solution to this problem.

Despite a steep decline in the contribution of agriculture to GDP, the labour force dependent on agriculture is still quite high, resulting in an increase in rural-urban differential in labour productivity and incomes. While the growth of rural non-farm enterprises alone provides an ultimate solution to this problem, the dependence of a large number of small farmers on agriculture for their livelihood will continue for quite some time to come. Owing to the continued pressure of population on limited land resources, tenancy has persisted through informal contracts. Because of the ceiling on land holdings, sub-division and persisting pressure on land as a source of livelihood, there has been an increase in the operated area - owned as well as leased-in - held by the small and marginal farmers. As expected, with the widespread adoption of new technology and rise in input-intensity, the incentives for investment and for capturing the returns on investment seem to be predominating over the need for sharing yield risks, leading to the gradual replacement of sharecropping tenancy by the fixed crop and cash rents.

Meeting the credit requirements of small and marginal farmers is crucial for their survival and for their ability to raise incomes by adopting new technology. Small farmers are in need of adequate *availability* of institutional credit rather than concessional rates of interest, as the rates charged by the institutional sources are significantly lower than those charged by the private sources. Experience shows that the repayment performance of small farmers has been as good as, or even better than, that of resourceful farmers. The exceptional performance of Self-Help Groups in respect of micro- credit is a case in point. Despite this, there has been a decline in the proportion of institutional credit made available to the small and marginal farmers in the post-reform period when their credit requirements have in fact increased on account of diversification and rising capital-intensity of agriculture. Innovative delivery systems on the pattern of group-lending have to be devised to provide institutional credit to small and marginal farmers. Exclusive emphasis on meeting the credit targets for agriculture sector *as such,* without any attempt towards sub-targeting the supply for small and marginal farmers, would only encourage re-lending by the resourceful farmers to the weaker sections at high rates of interest.

Rural Poverty Alleviation

The experience of the former socialist countries, including China, has shown that land reforms and infrastructure development, by themselves, would not be sufficient to eradicate poverty in a sustainable way if the productivity of resources is not raised by opening up the economy and increasing the role of markets for providing incentives to the producers. The post-reform experience in India shows, on the other hand, that opening up of the economy and increasing the role of markets, by themselves, would not be sufficient to make a dent on poverty unless conscious measures are undertaken by the state for rural infrastructure development and employment generation.

The direct impact of agricultural growth *as such* on the reduction of rural poverty in India through employment generation has been much less than desired, on account of widespread landlessness, regionally uneven development and sluggish rise in real wages in the less developed regions. Nevertheless, in the 1980s when agricultural growth picked up in the eastern region where poverty is extensive, the reduction in rural poverty there was higher than the national average. Even so, a direct attack on poverty became necessary by launching a number of Poverty Alleviation Programmes targeted at the poor. In general, wage employment programmes for Rural Works proved to be more effective than the Self-Employment Programmes based on the endowment of assets. This experience notwithstanding, there is a potential for gainful self-employment of the rural poor in the infrastructurally developed areas in activities that are land-saving, e.g., horticulture, dairying, poultry etc., the demand for which is quite buoyant. Likewise, wage employment programmes can be used productively for rural infrastructure development including eco-development such as soil and water conservation. This is how Poverty Alleviation Programmes can be integrated with overall development.

Improvement in the delivery systems is the key to the success of these programmes. The poor farmers have been able to successfully participate in green revolution because of the control they could exercise over their resources and because of the freedom for decision-making regarding the allocation of resources. Instead of being top-down programmes *for* the poor, as at present, the Poverty Alleviation Programmes need also to become programmes *of* the poor managed *by* the poor themselves through their elected institutions. While the reservations in jobs for backward classes and reservations in Panchayats for weaker sections, including women, have become mandatory, their *effective* empowerment is possible only by mustering the political will to devolve necessary *functions, finances and functionaries* to these elected institutions.

Involvement of NGOs and use of Information Technology can play a major role in raising the awareness of the poor, improving their capabilities and strengthening their group endeavour through necessary institution-building.

June 30, 2005. **C.H. Hanumantha Rao**

Institute of Economic Growth,
Delhi.

I

Agricultural Growth

1

Agricultural Performance and Emerging Perspectives

Introduction

Agriculture accounts for one-third of national income in India and about two-thirds of India's workforce is directly engaged in agriculture for its livelihood. It still accounts for sizeable export earnings. Nearly 80 per cent of India's poverty population (*i.e.*, population below the poverty line) lives in rural areas and is directly or indirectly dependent on agriculture.

India's water resources — both surface and ground water — when fully exploited, would bring half the cultivated area under irrigation. The abundant sunshine offers considerable scope for increasing crop yields as well as for harnessing forest resources as renewable sources of energy. The application of modern technology to agriculture is still very much lagging behind the advanced agricultural systems of the West as well as of the East. The successful application of such technology would, therefore, increase crop yields substantially. Thus agriculture in India offers considerable potential both for achieving self-reliance in food on a sustained basis and for eradicating poverty.

The growth of foodgrains output was negative during the 15 years prior to Independence. The colonial system perpetuated traditional technology in agriculture except in the case of a few export-oriented commercial crops. The development of infrastructure like irrigation, roads, markets etc., was extremely limited. The intermediaries in agriculture like *zamindars* and *jagirdars* had grown in the colonial period in large parts of the country, denuding the investible resources and sapping the motivation of the actual tillers of the soil. Thus, at the time of Independence, we inherited stagnant agriculture with deficient infrastructure, traditional technology and out-moded and exploitative institutions.

Thanks to the improvements in public health and the rapid reduction in infant mortality, we were faced with the problem of population growth outpacing the availability of food. The Five-Year Plans were, therefore, aimed at improving the infrastructure like irrigation, stepping up the use of fertilisers and improved varieties of seeds and effecting institutional changes, such as, the abolition of intermediaries, imposition of ceilings on land-holdings and the distribution of surplus land among the landless, supply of credit through the cooperatives, ensuring remunerative prices for agricultural commodities *etc.*

Institutional Context

It is important, at the very outset, to understand the institutional, demographic and socio-political context within which agricultural growth has been taking place. The institutional framework itself has undergone some changes while it has shown rigidities in certain respects. The prevailing institutional, demographic and socio-political framework has had a profound impact on the pattern of agricultural development. This in turn has necessitated special strategies and policies for employment generation and poverty removal in rural areas.

In the early years of planning, the strategy for employment generation and poverty removal rested on a series of expectations in regard to institutional changes such as land reforms. The two basic objectives of land reform were (a) to abolish intermediaries (such as *zamindars* and *jagirdars*) between the government and the tillers so as to ensure security of tenure to the latter and eventually to make them owners of land; and (b) to impose ceilings on the ownership of land holdings and distribute the surplus land among the landless poor. The intermediaries in land were abolished within a few years after independence and the actual tillers, accounting for about 40 per cent of cultivated area, became owners. This provided a major incentive for investment and growth of agriculture in large parts of the country. However, this reform did not meet with success in certain parts like North Bihar where erstwhile *zamindars* continued to own large holdings and lease out land to the sharecroppers.

As to the second objective of imposing ceilings on landholdings and distributing the surplus land among the landless, there were two rounds of land reform legislation — one in the fifties and another in early seventies. The State legislatures readily passed the radical legislation imposing ceiling on land holdings. However, due to numerous exemptions provided in law, and to the lack of political will, the objective of this reform was only partially fulfilled. The law discouraged concentration of land ownership beyond the ceilings level and thus prevented the possible dispossession of numerous small and marginal holders which would probably have occurred, through a competitive process in the land market in the absence of ceiling on landholdings.

Due to high population growth there was rapid sub-division of large landholdings and, in course of time, many of these holdings came well within the legal ceiling. On account of numerous exemptions from ceiling provided in law such as for plantations, and also due to the political dominance of large land owners, not much surplus land could be acquired by the government for distribution among the landless. In the course of the last four decades, not more than 2 per cent of total cultivated area in the country could be acquired as surplus for distribution among the landless. Thus the inequality in the distribution of land holdings persisted. This has lent considerable political clout for large farmers at various decision-making levels. This ranges from influencing agricultural policies at the government level to dominance in the institutions for the distribution of credit and inputs like modern varieties of seeds and fertilisers.

Thus while the abolition of intermediaries and ceiling on holdings have, no doubt, lent dynamism to agriculture and prevented the concentration of land holdings, the failure to redistribute land to the landless in any significant measures, has been a major cause for the persistence of rural poverty. Besides, as will be discussed later, this has resulted in a pattern of agricultural growth where the employment of hired labour (consisting essentially of landless poor) among the large farms has been slow in relation to the growth of agricultural output.

Agricultural Growth

It was becoming clear by the mid-sixties that there was no alternative to technological change in agriculture for achieving self-sufficiency in foodgrains. Even those countries in Asia which could carry out radical land reforms and build up adequate infrastructure for agriculture had taken to the path of modernising agriculture. Japan took the lead in this direction and China followed suit even after successfully experimenting with structural changes and mobilisation of growing labour force for capital construction in agriculture.

After the mid-sixties, apart from the introduction of HYVs of seeds for wheat and rice, public investment in agriculture was stepped up significantly. The new technology raised the profitability of investment for the farmers. Besides, in agriculture, there is a high complementarity between public and private investment. As a result, the annual gross capital formation in agriculture, public and private combined, at constant prices, rose much faster than in the pre-green revolution period. This is reflected in the accelerated increase in gross irrigated area. From less than one million hectares per annum during the pre-green revolution period, it increased to about 2.5 million hectares per annum during the seventies. However, owing to the land constraint, the growth of net sown area slowed down considerably. This explains why despite the accelerated increase in capital formation and the introduction of new technologies, the growth rate of agricultural output was no higher than in the pre-green revolution period.

Nevertheless, there was a rapid increase in the marketed surpluses of foodgrains and in the stocks held by the government through internal procurement. This is due to the basic changes in the structure of agricultural production witnessed in this period. First, the breakthrough in foodgrains output occurred in the developed regions in the North-West and in certain parts of the South. The consumption of foodgrains was already high in these areas and therefore, a considerably larger proportion of additions to output was marketed. Second, the use of labour per unit of output showed a significant decline, which also resulted in the generation of larger foodgrain surpluses from the rural areas.

There are two other factors responsible for the rise in marketed surpluses and availability of foodgrains. The increase in output during this period has been taking place mainly through the increase in yield per hectare. Therefore, there has been a decline in the ratio of seed to output, although our official estimates continue to be based on the conventional norms. Secondly, the rate of urbanisation has increased and the average per capita consumption of foodgrains in the urban areas has been significantly lower than in the rural areas.

We are used to comparing the performance in the pre-green revolution period with that in the post-green revolution period as a whole. However, the mid or the late seventies mark a definite departure from the trends observed in the early phase of green revolution in several respects. There seems to be some improvement in the growth rate of foodgrains output in the last decade, *i.e.,* 1978-79 to 1988-89, when compared with the first decade of green revolution. Crops such as rice, pulses and oilseeds, whose growth rates in the first decade of green revolution had fallen much below those recorded in the pre-green revolution period, are now showing higher growth rates. There are thus clear indications that the major inter-crop imbalances in growth witnessed in the early years of green revolution are getting redressed to some extent in the recent period.

Oilseeds and pulses are grown essentially in the rainfed or unirrigated areas. Even for rice, nearly 60 per cent of area is rainfed. The recent experience, therefore, suggests that the disparities in growth between the irrigated and rainfed or dry areas may not be as sharp as in the early years of green-revolution. It is also heartening to find that many of the States where poverty is widespread and where the growth of foodgrains output had slowed down in the first decade of green-revolution, *e.g.*, Assam, Bihar, Orissa, Madhya Pradesh and West Bengal have shown a much better performance in the last decade. Their performance is nearer, or even higher than, the all-India average. The growth of rice had slowed down considerably in the Eastern States during the seventies. All these States have experienced a step-up in the growth of rice during the last decade and some of them have recorded higher growth than the national average.

This encouraging performance is basically attributable to the spread of new technology to the lagging crops and regions. The special programmes for rice and oilseeds and increasing attention being given to dry farming and to the less developed States have started yielding results. This is corroborated by the increase in the consumption of fertilisers during the eighties at a faster rate in the Eastern and Western regions than in the Northern and Southern regions. What is more, fertilisers consumed in *kharif* season increased at a much faster rate than in *rabi* season, particularly in the Eastern and Western regions, indicating the spread of new technology to the rainfed and environmentally unfavourable areas. There is also evidence of an increase in the share of the Eastern States in the total gross irrigated area of the country.

However, the picture is not encouraging regarding capital formation in agriculture in the country during the eighties. After showing a steep rise in the seventies, the gross capital formation in agriculture at constant prices, both by the public and private sectors, showed a significant decline in the eighties. The decline in the net capital formation could be even steeper. While the gross irrigated area increased at the rate of nearly 2.5 million hectares per annum during the seventies, the increase in the eighties came down to less than a million hectares per annum, on an average.

The gains in foodgrains output in the last decade have come essentially from the improved utilisation of the available infrastructure and from the increase in yield per hectare. Already, during the first decade of green-revolution increase in yield per hectare accounted for as much as 80 per cent of the increase in foodgrains output. During the last decade, there was some decline in the area under foodgrains. The entire output growth in this period is, therefore, attributable to increase in yield per hectare. This underscores the growing importance of yield-increasing technology.

Productivity

Now, one can legitimately ask the question: At what cost has this growth been achieved? Indeed, based on the input data from the National Accounts Statistics, fears have been expressed that growth in agricultural output is being achieved with increasing costs per unit of output, as the index of inputs at constant prices has been rising faster than output (Rao, V.K.R.V., 1983). However, these data on inputs exclude land and labour. The new agricultural technology is essentially land-saving and also labour-saving to some extent. There is substitution of capital for land and labour. It is not surprising, therefore, that with the increasing application of new technology, the capital inputs should rise faster than output. What we should really be looking for is the productivity of total inputs — land, labour and capital combined.

The estimates of costs of production brought out by the Ministry of Food and Agriculture, Government of India, give the latest position in regard to the shares of land, labour and capital in the total costs. Applying these weights to the growth of different inputs, the index of total inputs is seen to be rising at a little less than 2 per cent per annum in the post-green revolution period against the growth in foodgrains output at 2.6 per cent per annum. On this basis, between one-fifth to one-fourth of the observed increase in the output of foodgrains can be attributed to the increase in the productivity of inputs. Commodity-wise, the contribution to the growth of output from the improved productivity of inputs may be nearly 45 per cent for wheat and about 15 per cent for rice (Tyagi, 1989).

Stability

Although the growth rate in foodgrains output seems to have improved because of the impact of new technology, the growth performance has by no means been smooth. The instability or the year-to-year variation in foodgrains output has increased in the post-green revolution period. Even within the post-green revolution period, the instability in the last decade (1978-79 to 1988-89) was higher than in the preceding decade. The rural poor, particularly in the drought-prone areas and in the remote areas of the country, continue to suffer from fluctuations in employment and income and inadequate availability of foodgrains in years of drought.

However, this increase in instability cannot be attributed to new technology. Rather, the instability arises from the adverse agro-climatic conditions in which the technology is used. The new seed-fertiliser technology has raised the response of output to water. Thus, for a given variability in rainfall or moisture conditions, the instability in output would be greater. However, when the new technology is applied under assured irrigated conditions, the increase in output would be on a stable path. This is exactly what has happened in the case of wheat where the new technology has made the maximum impact. The rise in instability is significant for crops like rice, oilseeds and pulses. This is explained by the increasing application of new seed-fertiliser technology for these crops in rainfed areas and under uncertain irrigation in the recent period. This suggests the need for stabilising consumption through larger procurement of foodgrains when harvests are good and their liberal release through public distribution system for the vulnerable groups in years of drought.

Equity

Perhaps the most widely debated issue about green-revolution is the growing disparities in income between different regions and classes of farmers. The experience after the mid-seventies covering nearly a decade and half reveals trends which are typical of a diffusion process. These trends are: the spread of green revolution to new areas, the increasing adoption of new technology by the small farmers, the decline in the relative prices of foodgrains and the rise in real wages in agriculture in the less developed regions where new technology is spreading.

Even so, the expectation that land-saving technological changes in a labour-abundant economy like India would lead to greater labour absorption in agriculture has been largely belied. The short-duration HYVs of seeds can facilitate an increase in cropping intensity and may thus lead to an increase in labour use. In practice, however, because of the profitability of intensive farming involving the use of large quantity of water and fertilisers per unit of land, there was not much incentive for increasing cropping intensity. Moreover, until recently, the impact of new technology and the expansion of tubewell irrigation was not significant in regions where labour is plentiful. This may explain why the cropping intensity has risen at a much slower rate in the post-green revolution period.

In the early phase of green revolution, large farmers, owing to better access to capital resources, stepped up yields per acre at a faster rate than small farmers. Because of this, in areas experiencing technological change, the inverse relationship between farm size and output per acre began to disappear. In course of time, however, the supply of institutional credit for the less developed regions and small farmers improved significantly (Desai, 1988; Haque and Verma, 1988). As a result of this and also because of improved extension services, the use of new seed-fertiliser technology among small farms caught up with that among large farms. And, because of the continued advantage that small farmers have in respect of cropping intensity, the inverse relationship between farm size and output per net operated acre has started reappearing (Bhalla and Chadha, 1983; Shergill, 1987). However, labour input per acre continued to decline sharply with the increase in the size of holding.

Insofar as large farmers have resumed land for self-cultivation from share-croppers because of the profitability of new technology, it has led to lower labour absorption and reduced incomes for erstwhile tenants. However, in quite a few places, cost-sharing on new inputs between landowners and share-croppers has become popular resulting in the sharing of benefits from new technology by the landowners as well as tenants.

Even in the early phase of green revolution, the emerging disparities between different regions were more conspicuous than the disparities between different classes of farmers within regions experiencing technological change. The uneven regional growth was mainly responsible for the low absorption of labour within agriculture. In a large number of States, especially in those regions where there was abundant availability of labour, the growth of output was too slow to generate adequate employment opportunities. In high growth regions labour was not plentiful and wage rates were high. The sudden rise in the demand for labour in these areas induced mechanisation and labour-saving practices in general. This happened despite the use of migrant labour from the less developed regions for certain operations. The net result was a significant decline in the elasticity of employment with respect to agricultural output for the country as a whole (Rao, 1975; Bhalla, 1987; Vaidyanathan, 1986). At the same time, the inter-State disparities in the growth of output and in output per worker continued to increase from the early-seventies through the early-eighties (Bhalla and Tyagi, 1989).

However, the inter-State disparities in agricultural wages started declining since the mid-seventies (Jose, 1988; Nayyar, 1987). The decline in disparities in real wages is even more significant. This is because the decline in the relative prices of foodgrains had a greater impact on the purchasing power of wage earners in low wage areas. For example, the disparities in real wages in agriculture between the Punjab, on the one hand and Bihar, on the other, started coming down since the mid-seventies. The real wages in the Punjab having risen significantly in the early phase of green revolution, virtually stagnated after the mid-seventies whereas the real wages in Bihar started rising. The cause for the decline in inter-State disparities in real wages are: the out-migration of labour from the less developed regions to the high wage areas; mechanisation of agricultural operations in the developed regions; the decline in the relative prices of foodgrains; the increase in employment generated in the less developed regions under the poverty alleviation programmes; and the recent pick-up in agricultural growth in these regions. An encouraging consequence of these developments is that the disparity between male-female wage rate has started declining in several States. The use of new technology may have improved the bargaining power of female labour for operations such as transplanting and inter-culturing on account of the rise in seasonal demand for such labour.

Low income groups, both in the rural and urban areas, would derive greater benefit from the decline in the relative prices of foodgrains, because they spend a larger proportion of their income on foodgrains than the upper income groups. Agricultural price policy has been able to ensure that the productivity gains from technological change are shared both by the consumers and producers. However, consumers seem to have benefitted relatively more than the producers from the productivity gains, particularly in the case of rice and *jowar*. The demand for wheat has been more favourable than for other cereals. So the decline in the relative price of wheat was only about 25 per cent, while the productivity gains amounted to nearly 45 per cent of output growth.

Thus the indirect contribution of green revolution to equity through food security and the decline in the relative prices of foodgrains has perhaps been more significant than its direct impact by way of labour absorption in agriculture. Because of slow growth of employment in agriculture, the employment and income generated under poverty alleviation programmes had to be stepped-up by using surplus stocks of foodgrains. From about 20 million man-days of employment generated annually in the mid-sixties, the employment generated under such programmes in the country as a whole amounted to 850 million man-days in 1988-89. These employment programmes together with the income generated under the Integrated Rural Development Programme (IRDP) seem to make up for about half the deficiency in employment generation in agriculture in the post-green revolution period. The relative contribution of these poverty alleviation programmes to employment generation is more significant in the poorer regions where agricultural growth has been below the national average. These programmes have been made possible because of the increased availability of foodgrains from internal procurement.

Sustainability

Unlike in the intensive systems of agriculture where pollution of environment has been caused by the intensive use of chemical inputs, the ecological degradation in rural India is attributable mainly to the extension of cultivation to the marginal and sub-marginal dryland and to deforestation. In India, the proportion of rural population below the poverty line is highest for unirrigated or dry areas and declines with the increase in irrigation (Rao, Gupta and Sharma, 1986). Therefore, the degradation of the already fragile environment in dry areas has further accentuated the poverty problem.

The causes of deforestation and degradation of dry lands are many. Apart from population growth and poverty, the pressure of demand for fuel wood, timber, meat and wool from high income segments including urban areas has been a major cause of denudation of environment. However, the rural poor have been the prime victims of this ecological degradation. This is because, the rural poor derive a significant part of their income from village commons and forests in the dry regions (Jodha, 1986). Even though the proportion of income derived from such sources is greater for the rural poor than for the non-poor, the latter are responsible for the greater share of denudation of forests than the rural poor (Nadkarni, 1989). As a result of environmental degradation, the rural poor, particularly women and children, have to walk much longer distances to collect fuel wood and fodder, and drinking water has become scarcer owing to the decline in water table.

There is a clear inter-relationship between the levels of agricultural development and rural poverty on the one hand, and ecological degradation on the other (Mellor, 1988). Across different States in India, the extension of area under cultivation and the denudation of forests seem to be high where the progress of yield-increasing technology is slow. In such regions, the levels of agricultural income and wages are low and poverty levels are high. Similarly, the pressure from animals such as goats and sheep on forests and common lands has been increasing in regions where growth in crop production is slow. This is because the rural poor supplement their incomes by rearing these animals (Rao, 1991).

This experience suggests that apart from yield-increasing technology for crops, the use of new bio-technologies in forestry and animal husbandry to raise the yields from trees and animals and improvement in management practices may contribute significantly to the protection of environment as well as to raising the wages and incomes for the rural poor.

Agricultural Price Policy

In the pre-green revolution period, owing to the persistence of severe imbalances between the domestic demand for and supply of agricultural commodities, particularly foodgrains, the terms of trade in general turned favourable to agriculture. Agricultural price policy was concerned in this period mainly with the procurement of part of the foodgrain surpluses from the farmers at prices lower than the free market prices with a view to protecting the consumers, especially in the urban areas.

In the post-green revolution period, however, agricultural price policy became increasingly producer-oriented. The proportion of marketable surplus of wheat and rice procured from the farmers by the public agencies, at pre-determined procurement prices, increased steadily which helped to prevent a dip in market prices in years of good harvests. With the progress of green revolution and the concentration of surpluses in a few regions, the need to protect farmers through such a policy became increasingly important. Despite this, the terms of trade for agriculture became slightly unfavourable during the eighties, indicating that part of the benefits from rising productivity on account of technological change have been passed on to the consumers. This was particularly so in the case of rice and wheat, the demand for which is relatively inelastic.

The Commission on Agricultural Costs and Prices constituted in mid-sixties continued to provide expert advice to the government on the prices to be fixed in the light of the prevailing cost of production, variations in supplies in relation to demand and overall price movements in the country as well as in world markets. In practice, the prices fixed by the government were in quite a few cases higher than those recommended by the Commission. In a developing economy like India, the non-price measures for raising agricultural output such as provision of basic infrastructure like irrigation and power, and extension of new technology by making available to the farmers high yielding seeds and fertilisers are far more important than incentive prices. Also, in a democratic policy faced with growing population and widespread poverty, the need to supply essential commodities like foodgrains to the consumers at affordable prices assumes high priority. It is fair to conclude, therefore, that agricultural price policy in India has, by and large, tried to balance the interests of producers and consumers in the light of overall goals of development.

Agricultural Taxation

The burden of land revenue was quite heavy during the colonial period, very often accounting for half the net income from agriculture. The surpluses so mobilised were invested, among others, in major irrigation systems, railways, power generation, urban development *etc*. The heavy taxation of agriculture together with the prevalence of outmoded land tenure systems denuded the capital base of agriculture. After independence, the need to meet the food requirements of growing population necessitated sizeable public investments in agriculture. The rise of agricultural classes to political power at various levels through democratic processes facilitated a significant resource inflow into agriculture. Due to the steep rise in the prices of agricultural commodities and land revenue rates remaining constant for several decades, the impact of land revenue declined steeply to below 1 per cent of net income from agriculture.

The issue of taxing agricultural incomes on par with urban incomes has been debated extensively during the plan period. However, the existence of legal ceilings on landholdings and the continued subdivision of landholdings effectively limits the scope for mobilising revenues from taxing agriculture. Due to the well-known limitations of arriving at dependable estimates of net income of farm households and the possibility of widespread evasion, the Expert Committee headed by K.N. Raj had recommended progressive taxation of agricultural land in the early-seventies.

Although the net revenue realisation after accounting for the costs of administering such a tax system may not be significant, there is still a case for taxing upper income groups in agriculture in the interests of inter-sectoral as well as intra-sectoral equity, particularly in areas experiencing green revolution. Such a direct taxation of agriculture has, however, proved to be politically infeasible. An alternative form of taxing agriculture through a levy on surpluses marketed by the farmers at the market yard, experimented in Punjab in the recent period, has already brought in sizeable revenues. The incidence of such a levy would be progressive because the marketed surplus as a proportion to output increases significantly with the rise in the size of holding.

Input Subsidies

Selective subsidies on technologically new inputs like fertilisers are generally preferred to the rise in output prices for promoting the adoption of new technology, particularly because such subsidies can be targeted to the small farmers and the lagging regions. The subsidy on fertilisers did contribute to the widespread adoption of new technology in Indian agriculture. However, over a period of time, the subsidy on fertilisers as well as implicit subsidies on non-tradable inputs like irrigation and power have assumed enormous proportions and are no longer justified purely on promotional grounds.

The annual subsidy on fertilisers alone amounts to nearly as much as the annual plan outlay on agriculture by the Centre and States put together. Fertiliser industry, farmers and consumers at large benefit from these subsidies. Among farmers, a disproportionately large part of the benefit accrues to those in the prosperous irrigated regions which have taken a lead in the adoption of new technology. The implicit subsidies on irrigation and power arising from the non-recovery of economic rates significantly exceed the explicit subsidy on fertilisers.

These subsidies have had adverse macro economic consequences insofar as they have caused large budgetary deficits. The recent trend of a decline in real investment in agriculture is traceable to the paucity of investible resources caused, in no small measure, by the rise in subsidies. These subsidies have induced greater allocation of resources to irrigated crops like rice, wheat, cotton and sugarcane at the expense of crops like oilseeds and pulses the prices of which have been rising steeply. Further, over-use of canal water in the head reaches and overdrawal of groundwater have led to environmental problems such as water logging on the one hand and severity of droughts on the other.

Thus whereas input subsidies have undoubtedly contributed to the widespread use of new technology and achievement of self-sufficiency in foodgrains and benefitted the consumers by bringing down the relative prices of foodgrains, the subsidies have reached a point where their beneficial effects seem to be outweighed by the adverse effects in terms of macro-economic imbalances, slowing down of investment in agriculture, inefficient use of resources at the farm level and adverse impact on environment. There is, therefore, a need to reduce the quantum of these subsidies and target them, to the extent possible, to the small farmers and the lagging regions. Since there is now an overuse of these inputs in limited pockets, the move towards efficient pricing of such inputs may lead to greater diversification of their use benefiting larger number of crops spread over wider areas and may, in general, result in the rise in output per unit of inputs used. Vulnerable sections of consumers can be adequately protected by strictly targeting food subsidies to such groups.

Trade Policies

India's share in the world exports of agricultural commodities has been declining sharply. Whereas there is a potential for raising India's share in world exports of agricultural commodities, this potential can be realised only by raising agricultural productivity so as to augment supplies in relation to domestic demand. Investment and technological change in agriculture hold the key to raising productivity. Restrictions on exports have been imposed from time to time to protect the interests of domestic industry as well as consumers by stabilising prices. Imports of certain commodities like edible oils are also restricted for conserving foreign exchange as well as for ensuring incentive prices to the farmers.

Government intervention in trade seems to have led to some inefficiencies, particularly in the eighties, when, as a result of technological change and heavy input subsidies, foodgrain stocks with the government have been considerably augmented especially in years of good harvests, and the internal carry-over costs have increased. Wheat, rice and cotton have emerged as good exportables upto a point, particularly after the recent adjustment in the exchange rate, as they have been 'disprotected' whereas imports of edible oils, which have been heavily protected, could be increased (Gulati and Sharma, 1991). However, as pointed out earlier, if input subsidies are brought down, there would be some diversion of resources from water-intensive and fertiliser-intensive crops to semi-irrigated crops with a consequent change in the relative prices of crops, thus narrowing down the difference between the international and domestic prices of cereals as well as oilseeds. Thus a good part of the 'distortions' attributed to restrictions on trade are indeed the result of domestic distortions arising from heavy subsidies on inputs.

Even so, there may be some gains through trade by specialising in crops like cereals and cotton and importing edible oils upto a point. This may increase employment opportunities in the country as the exportable crops are labour-intensive. However, a few qualifications are in order. World trade in grain is highly volatile and considering the possible impact of India's decision to export grain on world prices, exports may cease to be profitable beyond a point. Secondly, in view of the uncertainties of world supplies, the country can not run the risk of undermining its food security by running down the domestic stocks below a certain critical minimum considering the incidence of droughts and commitments of the public distribution system to meet the requirements of a large population still living below the poverty line. Also, care has to be taken to ensure that the prospects for technological change in respect of oilseeds, if any, are not jeopardised because of unremunerative prices resulting from heavy imports.

However, broadening the domestic agricultural base by stepping-up public investment in infrastructure, research and extension is essential for exploiting the possible gains from trade as well as for ensuring that such gains are shared widely by different regions and classes of farmers. Simultaneously, it would be necessary to protect the vulnerable sections of population through the subsidy on the sale of foodgrains. Such subsidies may indeed be justified upto a point by the gains to the society from trade.

Conclusion

The on-going economic reforms in the country have direct as well as indirect bearing on the prospects for agriculture. The most important consequence of the reduction in input subsidies for agriculture and relaxation of restrictions on trade in agricultural commodities would be to raise the investible resources with the government as well as the farmers. This tendency would be reinforced in the long-run because agriculture would also benefit significantly from the improvements in the domestic terms of trade, as industrial inputs for agriculture and consumer goods become competitive and cheaper consequent to deregulation and opening up of the economy. This is because domestic industry has so far enjoyed a high rate of protection while agriculture has, on the whole, been disprotected (Gulati and Pursell, 1991).

The need for stepping up investments in agriculture is underlined by the recent trend of a decline in real investment in this sector. Also, there is a need to cope with some of the adverse consequences of economic reform in the short-run such as inflation and slow growth in output and employment. Agriculture can work as the biggest safety net in the process of adjustment by softening the rigours of inflation as well as by raising income and employment for the vulnerable sections of population. Stepping-up public investment in essential infrastructure like irrigation, research and extension and in social development such as rural health, education and skills holds the key to realisation of the potential for private investment and effort.

References

Bhalla, G.S. and G.K. Chadha (1983). *Green Revolution and the Small Peasant*, Concept Publishing Company, New Delhi.

Bhalla, G.S. and D.S. Tyagi (1989). *Patterns in Agricultural Development: A District Level Study*, Institute for Studies in Industrial Development, New Delhi.

Bhalla, Sheila (1987). "Trends in Employment in Indian Agriculture, Land and Asset Distribution," *Indian Journal of Agricultural Economics*, Vol.42, No.4, October-December.

Desai, D.K. (1988). "Institutional Credit Requirements for Agricultural Production-2000 A.D.," *Indian Journal of Agricultural Economics*, Vol.43, No.3, July-September.

Government of India, (1989). *National Accounts Statistics 1980-81 to 1986-87* (New Series), Central Statistical Organisation, Department of Statistics, Ministry of Planning, New Delhi.

Gulati, Ashok and Sharma, Pradeep K. (1991). "Government Intervention in Agricultural Markets: Nature, Impact and Implications," *Journal of Indian School of Political Economy*, Vol.3 No. 2, April-June.

Gulati, Ashok and Pursell, Garry (1991). *Trade Policies, Incentives and Resource Allocation in Indian Agriculture*, (mimeographed), World Bank.

Haque, T. and Sunita Verma (1988). "Regional and Class Disparities in the Flow of Agricultural Credit in India," *Indian Journal of Agricultural Economics*, Vol.43, No.3, July-September.

Jose, A.V. (1988). "Agricultural Wages in India," *Economic and Political Weekly*, Vol.23, No. 26, June 25.

Jodha, N.S. (1989). "Dry Farming Research: Issues and Approaches," in N.S. Jodha (Ed.) (1989). *Technology Options and Economic Policy for Dryland Agriculture: Potential and Challenge*, The Indian Society of Agricultural Economics, Bombay; Concept Publishing Company, New Delhi.

Mellor, John W. (1988). "The Intertwining of Environmental Problems and Poverty," *Environment*, November.

Nadkarni, M.V. (1988). "Crisis of Increasing Costs in Agriculture: Is there a Way Out?," *Economic and Political Weekly*, Vol.23, No.39, September 24.

Nayyar, Rohini (1987). "Female Participation Rates in Rural India," *Economic and Political Weekly*, Vol.22, No. 51, December 19.

Rao, C.H. Hanumantha (1975). *Technological Change and the Distribution of Gains in Indian Agriculture*, The Macmillan Company of India Ltd., Delhi.

———. (1990). "Some Inter-relationships between Agricultural Technology, Livestock Economy, Rural Poverty and Environment: An Inter-State Analysis for India," in *Golden Jubilee Volume: Agricultural Development Policy: Adjustments and Reorientation*, Indian Society of Agricultural Economics, Bombay; Oxford & IBH Publishing Co. Pvt. Ltd., New Delhi.

Rao, V.K.R.V. (1983). *India's National Income, 1950-1980: An Analysis of Economic Growth and Change*, Sage Publications India Pvt. Ltd., New Delhi.

Shergill, H.S. (1987). *Impact of New Technology on Farm Size-Productivity Relationship in Punjab Agriculture: A Decomposition Analysis*, Department of Economics, Punjab University, Chandigarh (mimeo).

Tyagi, D.S. (1990). *Managing India's Food Economy: Alternatives in a Self-Sufficiency Environment*, Sage Publications, New Delhi.

Vaidyanathan, A. (1986). "Labour Use in Rural India: A Study of Spatial and Temporal Variations," *Economic and Political Weekly*, Vol.21, No.52, December 27.

Published earlier in *The Indian Economy: Problems and Prospects*, Bimal Jalan (ed.), Viking, New Delhi, 1992.

2

Agricultural Policy under the Plans

Introduction

Agricultural policies can be understood and evaluated only with reference to the overall strategy of economic development and the basic policies pursued to achieve the goals set for the country. This in turn involves an understanding and appreciation of the existing social structure and the political framework which are given as basic data and within which all economic programmes have to be drawn and implemented.

The basic objectives as well as the processes of economic planning in India can be summed up as follows: A rapid rate of economic growth with increasing emphasis on the welfare for the masses of people. These two objectives have become historical imperatives or inevitables for India. There is no alternative to the rapid rate of economic growth for a country with widespread poverty, population explosion and the rising expectations. This necessarily involves the active participation of the state, which alone can combine social perspective or vision with required resources and discipline for undertaking those economic activities which are beyond the means of private individuals as well as for directing the economy as a whole. That all this has to be achieved through democratic methods is dictated by the prevailing social structure in India characterised by the co-existence of highly stratified and non-polarised class and caste groups and interests which considerably overlap each other. The second objective *viz*., the welfare of the masses needs to be spelt out explicitly. The objective in this regard as incorporated in the Indian plans is the gradual reduction in the inequalities in incomes between different groups as well as regions. But what is crucial to this process of growth is not whether the degree of inequality actually decreases over time but that there should be a sustained and reasonable increase in the standards of living of the poor – which cannot be postponed in a poverty-ridden and awakened society and which can be achieved despite increasing inequalities in incomes.

Obviously, this model is a via-media between the two extremes: a very high rate of growth with comprehensive planning under a relatively monolithic and authoritarian framework on the one hand and the *laissez-faire* on the other, which is chaotic, slow-moving; and, therefore, vulnerable to violent upheavals. The rest of this paper examines whether, and how far, our agricultural policies fitted into this framework. For a proper understanding and evaluation of agricultural policies it is essential to know the social composition and the ideological predilections of the policymakers. This is attempted in the following section. We have had a bunch of policies, successful as well as unsuccessful and those potentially useful but adopted with a considerable lag. Many of them did not fit into the basic objectives stated above nor were they always in harmony with each other. Yet, many of the achievements and failures on the agricultural front may have had very little to do with the agricultural policies of the Government. It is, therefore, important to get a rough idea as to the magnitude of agricultural performance which is autonomous in the above sense, for this would give a more realistic picture of the contribution of conscious policies to agricultural development. An attempt is made, therefore, to assess the contribution of such autonomous forces in the next section. Agricultural policies are then evaluated grouping them broadly into two, *viz.*, successful or productive and those which failed. Policy problems concerning future which are still shrouded in darkness and uncertainty but which may deserve the highest attention of planners and policymakers are discussed in the end.

Policymakers: Who's Who?

In theory, the policies formulated by a government constituted through the popular vote reflect the consensus of the community as a whole. This may be true in practice too, especially of those policies towards which public opinion is particularly sensitive. However, in the framing of policies which embody innovations and experiments such as in the field of economic development, the understanding and persuasion of the leadership may be more important than the immediate popular consensus behind such policies. Moreover, in regard to policies affecting particular economic sectors, a consensus among the policymakers and the legislators does not necessarily imply a favourable response from the sectors concerned. Therefore, for a proper understanding of the policies for agricultural development it is necessary to know the background and preconceptions of those who are actively engaged in policymaking.

The three architects of economic policymaking in India in the descending order of their role are the Politician, the Civil Servant and the Expert — mainly the economist. Ever since Independence there has been a noticeable cleavage between the political leadership at the Centre and State levels regarding their approach towards agriculture. The Central leadership has been strongly urban-oriented in its background but had definite views about the strategy for agriculture which the leadership had been contemplating and working out even long before Independence. Their attitude towards agriculture has essentially been one of drastic reform and reconstruction on the basis of several schemes floated and pampered from above. The hold of national leadership over the peasant masses and the intelligentsia was so strong that the opposition to their schemes within the ruling party was not effective even when many of these schemes could not be implemented and virtually remained on paper. This prevented a proper flow of genuine information from below to the top leadership which is so essential for the formulation and implementation of realistic programmes. This is especially serious because the number of intermediaries involved between the peasant and the highest policymakers is much larger than in any other economic sector. Although the rural vote is predominant in the country, the proportion of members with rural background was much less in the first Parliament. However, the weight of rural members has been increasing steadily in the successive Parliaments.

The leadership at the State level, on the other hand, has been composed of rich landed sections from the very beginning, whose hold on State affairs continued to increase over this period. Unlike the *zamindars* and the *jagirdars* they were rooted in the villages and displayed considerable drive for progressive cultivation. They had considerable influence over the peasantry and constituted the local base for the central leadership before Independence. However, they neither had the courage nor felt the need to openly oppose central leadership on several schemes of agrarian reform and reconstruction. They in fact passed legislations in favour of some of these measures but saw to it that they became infructuous in actual implementation. Agriculture being a State subject they have been in a position to undermine, if not reject, schemes which went against their interests and mobilise the State power and resources to subserve their interests.

The civil servant provided a continuity of policy administration between the pre-independence past and the post-independence period. The civil servant's impact on agricultural policy reflects his attitude towards the peasant or the trader as a reformer and the disciplinarian, on the one hand, and the benefactor or saviour, on the other. These attitudes were inherited from his role in the pre-independence period. He has tended to regard the Indian cultivator as basically irrational *i.e.*, ignorant of, and, therefore, very often acting against his own best interests. The farmer is regarded as, by and large, allergic to change even when it benefits him. Also, all too often he suspects the farmer and the trader of anti-social and anti-government behaviour. In his role as the saviour and benefactor he finds it prudent to make basic decisions himself for the farmer and uplift the latter through ingeneous schemes floated and administered from above with necessary financial support.

These attitudes as well as the expertise of the civil servant no doubt facilitated the undertaking and execution of a number of worthwhile reform measures as well as constructive projects for agriculture. But these attitudes may also mean the adoption of policies which reflect an insufficient appreciation of the role of market mechanism, the absence of respect for the farmer and consequently the lack of confidence in farmer's own initiative and resourcefulness, and indifference towards the sterile effects of pampering the farmers through doles and subsidies from above. Moreover, rooted as the civil servant is in secure urban environments, the distance between him and the farmer is very great and rarely has he any opportunity for a genuine comprehension of the requirements and urges of the farmers.

The progress of economic planning in India has been characterised by the increasing association of economists — Indian as well as foreign — with the processes of policymaking. These economists have, by and large, displayed a strong commitment to the objective of a rapid rate of growth of Indian economy. They have naturally been fascinated with the course and processes of economic development in countries showing very high rate of growth in a relatively short period *e.g.*, Japan,

Soviet Union and China. They systematise with a considerable degree of sophistication the relationships between the key economic variables in these high-growth economies and sharpen their models as much as possible by ignoring the institutional and political framework within which such a high rate of growth was in fact achieved. These models — largely abstracted from institutional and political constraints — are then used to prescribe economic policies for the growth of developing countries like India. Many of these economists reveal an open aversion to the possible relevance of non-economic factors, the discussion of which is unexciting to them as this would mean the sacrifice of 'sophistication' and 'exactness'. As a consequence, the policy preferences for Indian agriculture revealed by these economists are invariably those which are appropriate to, and can only be implemented in, situations with an entirely different socio-political framework from that obtaining in India.

A typical growth-economist looks towards agriculture for finding surpluses for economic development. Agriculture thus interests him mainly as a source of cheap labour, marketable surplus — food, industrial raw materials and exports — and tax revenues. His accent is on *mobilising* such resources from agriculture through appropriate institutional devices rather than on *generating* them through significant increases in agricultural productivity, as the latter would involve the diversion of scarce investible resources from the 'priority' sectors. He, therefore, becomes pessimistic whenever the terms of trade turn in favour of agriculture and feels encouraged about growth prospects whenever these are turned against agriculture. He seldom regards agricultural sector *as such* as a major source of national income growth or of general welfare. Consequently, he evinces minimal interest in the problems of capital formation in agriculture or of its modernisation. Like the civil servant he views the peasant with suspicion whom he equates with the '*kulak*' — an unavoidable evil who should be made to behave himself. Such an attitude towards agriculture on the part of a typical academic economist should not be surprising in view of his essentially urban middle-class background and exotic training in growth economics, which has tended to over-emphasise the technological and economic relationships to the relative neglect of the institutional setting within which economic growth is to be brought about.

This gap could perhaps have been bridged somewhat if sociologists had been actively associated with the planning processes. Social anthropologists in India have not made any serious and convincing attempt so far to identify the impact of the institutions like caste, joint-family, religion *etc*., on the preferences for, and the processes of, economic endeavour and to scrutinise the current policies for socio-economic development in the light of such knowledge. In particular, they rarely identify those social institutions and attitudes which may be congenial to economic growth and which can be exploited for this purpose. There is no doubt, however, that economic policies in India would gain in realism in the measure in which such insights are brought to bear on policy-making by the active association of growth-minded sociologists at the highest policymaking levels.

Contribution of Autonomous Factors

Land and labour (human as well as bullock) are the principal sources of output growth in traditional agriculture. Provision of irrigation water from wells and tanks by the use of human and bullock power and farm-yard manures also belong to this category. All these sources may broadly be categorised as autonomous in the sense that they are provided by the cultivators from the agricultural sector itself. On the demand side the growth of population is a major autonomous factor apart from urbanisation and rise in per capita income which may have been induced essentially by the process of planned economic development. However, none of these factors can be treated as completely autonomous because policies concerning land reform, prices and credit can affect the level and the intensity of use of traditional inputs as well as the demand for agricultural commodities. Public irrigation projects, fertilisers and the new technical knowledge may be regarded as induced factors on which the influence of public policies is overwhelming. Land reform, credit, price and tax policies may also belong to this group, although, as argued later, some of these may have been induced by the autonomous factors. It is difficult to quantify precisely the contribution of each of these categories of factors to the growth of agricultural output. However, it is possible to make an assessment in general terms especially regarding the changing significance of these two groups of factors over the plan period.

The two major factors accounting for spontaneity in agricultural performance in the post-independence period are the growth of population and the rise of agricultural classes to political power. Population growth has meant an ever increasing agricultural labour force which contributed to a better exploitation of traditional techniques of farming. Culturable waste lands held by the government were distributed to the landless labourers in many parts of the country. In many areas people on their own encroached upon waste lands for cultivation. Lands hiterto left fallow and uncultivated by large land owners were brought under plough in the wake of increasing demand for agricultural commodities. Among land-holding classes, the increase of family labour led to the sub-division of holdings and to the intensive cultivation of their land through the greater use of labour for irrigation and better crop pattern *etc*.

Political Independence released the initiative of agricultural classes, especially the richer ones, who came to wield considerable political influence and power at the district and State levels. This meant an increasing use of State machinery and resources for agricultural betterment. This is reflected, among other things, in the significant expansion of credit facilities for agriculture in the post-independence period. In this sense, some of the government policies and measures cease to be wholly exogenous to the agricultural sector, as they would have been carried through under the local pressures even in the absence of a concerted plan of development. The impact of these spontaneous forces extends much further. As argued later, although the influence of these forces on the basic policymaking was not significant in the initial stages of planning, their impact continued to increase and was decisive in the adoption of the new strategy of agricultural development towards the end of Third Plan.

The expansion of output through the fuller exploitation of traditional techniques of farming was very much induced and sustained by the growing demand for agricultural commodities. This was reflected in the generally favourable terms of trade that agriculture enjoyed throughout the plan period. This is obviously an exogenous factor so far as agricultural sector is concerned and may be traceable ultimately to the policies and measures of overall economic growth resulting in the increased demand for agricultural commodities. However, this particular stimulus to agricultural performance cannot be traced to any specific policy measure calculated to promote agricultural development.

It would thus appear that agricultural growth was significant so long as the autonomous factors were important and it stagnated with the exhaustion of these sources. It is in this later stage when agricultural policy came into its own as a major variable for increasing agricultural productivity through the provision of necessary modern inputs that the failures on agricultural front became conspicuous.

Agricultural Policy: Areas of Achievement

The most significant achievements of agricultural policy in the plan period are: (i) the abolition of intermediaries, *viz.*, *Jagirdari* and *Zamindari* system; (ii) the association of agricultural classes with the formulation and execution of development programmes at the local level under the scheme of democratic decentralisation; (iii) provision of irrigation through major and medium irrigation projects and of power for minor irrigation through rural electrification; and (iv) the expansion of credit through cooperatives. The first two measures represent a major institutional reform effected in independent India and the other two constitute the core of infrastructure which are beyond the means of individual cultivators and which could be provided only through the active participation and direction of the government. It is interesting to note that a firm start in respect of all these measures was already made in the early years of the First Five Year Plan which were continued in the subsequent plans.

The policy of agrarian reform in India concerned with action on three fronts: (i) the abolition of intermediary tenures like *zamindaris*, *jagirs* and *inams* which covered more than 40 per cent of the area of the country; (ii) the security of tenure to tenants in *ryotwari* areas and the regulation of rents; and (iii) the imposition of ceiling on land-holdings and the distribution of surplus lands to the landless and uneconomic holders. The policies in respect of the latter two failed, by and large, and are discussed in the next section. It is in respect of the first that success has been almost complete.

The abolition of these intermediary tenures brought more than 20 millions of tenants into direct relationship with the State. It was much more than an economic measure. Many of these *zamindaris* and *jagirs* had their own revenue administrations and for all practical purposes functioned as states within the states. The cultivators were burdened with several feudal dues and had to live under repressive and outmoded administrations. Their abolition and the reform of revenue administration in these areas not only gave adequate incentives to the cultivators for progressive cultivation but released their initiative on the cultural and political plane as well. Many of these *zamindars* and *jagirdars* were allowed to resume lands for self-cultivation which no doubt led to the eviction of tenants to some extent but contributed to the active participation of erstwhile landlords in farming. Although, part of these resumed lands may have been under-utilised or left fallow, the abolition of this system brought considerable areas of cultivable wasteland under the management of the government which were distributed to the landless contributing to the increase in area under cultivation in the plan period.

Although the scheme of democratic decentralisation (*Panchayati Raj*) was extended throughout the country towards the beginning of the Third Plan, the basis for it was already laid down with the first general elections when the rising agricultural classes began to wield political power at the local level. Their influence continued to increase in the subsequent period and the Community Development administration was faced with the problem of coping with them. The scheme of democratic decentralisation ended this confusion and conflict by subordinating the Community Development administration to the rural leadership. The scheme no doubt brought the caste-based as well as political factions to the forefront and in general provided the means of satisfying the age-old hunger for power by the rural elite. While this sharing of power can not be equated with their participation in development programmes, the experience of the Community Development Programme made it clear that it was difficult to secure people's participation in development without entrusting them with the responsibility for formulation and administration of the programmes. The prospects of development seemed better with such an integration than with continued conflict between the Community Development officials and the political elite.

With the agriculturists assuming power, the transmission of technical knowledge to the farms from the urban-based experts became relatively quick and the supply lines for inputs became more elastic. The major bottlenecks now lay in the quality of knowledge transmitted and the quantities of inputs supplied, both of which depended on the overall strategy of agricultural development. A consequence of democratic decentralisation which has been amply demonstrated, though not conclusively proved, is the increasing share of benefits going to the few richer sections of cultivators. This can be rectified only through the improvement in the bargaining power of the lower classes which can be expected in course of time with the spread of literacy and education and with increasing political awareness of the masses.

In regard to the provision of infrastructure, *e.g.*, irrigation and credit the achievements are substantial, although they can not be equated with success considering the requirements of Indian agriculture. Total area irrigated in the country registered an increase of 36 per cent in the first ten years of planning. Although the achievement in respect of major and medium irrigation sources fell considerably short of targets during the Third Plan period, the targets relating to minor irrigation were fully achieved. All this represented an increase of about 25 per cent in the irrigated area during the Third Plan period. The progress in minor irrigation reflects increasing mobilisation of local resources for long-term investments. This process has been very much facilitated by the public programmes of rural electrification. The number of towns and villages electrified represented a twelve-fold increase between 1951 and 1966. The extension of power in this period has been essentially to villages and small towns. There has thus been a recognisable improvement in the capital base of agriculture in per capita terms during 15 years of planning.

Since area under cultivation has increased substantially during this period, the proportion of area irrigated increased only marginally from about 17 per cent at the beginning of the First Plan to about 20 per cent towards the close of the Third Plan. Since about half of this irrigated area is represented by minor irrigation sources which are essentially dependent on monsoons, it would follow that as much as 90 per cent of the cultivated area in the country continues to be vulnerable to vagaries of monsoons even after 15 years of planning. The situation may well be far more serious because one can not place total reliance on the statistics regarding the achievements in respect of minor irrigation owing to the reporting biases on the part of the farmers as well as the agencies implementing these programmes.

Short and Medium term credit supplied to farmers through cooperative agencies registered an eleven-fold increase between 1950-51 and 1962-63. As a result, cooperatives accounted for about 26 per cent of credit supplied to farmers in 1961-62 as against only about 3 per cent a decade earlier. However, bulk of the credit is still supplied by the private sources, of whom professional moneylenders' share is predominant. Moreover, a significant portion of the credit from cooperatives is appropriated by the agricultural moneylenders whose lending to the farmers has increased very much during this period. Also, the position is far from satisfactory regarding the repayment of cooperative loans as well as the purposes for which these loans are used.

Agricultural Policy: Areas of Inconsistency and Failure

The most serious deficiency in respect of agricultural performance has been the continued imbalance in the growth of this sector relative to the rest of economy. The First Five Year Plan was essentially directed towards rehabilitation and consolidation of the economy with a clear emphasis on agricultural programmes. The Second Five Year Plan actually marked the beginning of a serious and concerted drive for economic development and it is precisely in this period that agricultural policies betrayed a doctrinal approach revealing inconsistencies between the accepted goals and the concrete measures.

As argued earlier, the emphasis on certain basic and heavy industries is an imperative for a country like India. However, it was neither imperative nor desirable to follow the socialist countries in regard to the techniques of keeping agriculture and the rest of the economy in balance. In these countries, the accent has been on the *mobilisation* of agricultural surpluses through cooperatives and collectives and by taxing agriculture heavily or by turning the terms of trade against it. Unlike these countries, we were embarking upon the road for planned economic development with a much poorer agricultural base, rapidly growing population and a socio-political framework which ruled out concentration of decision-making and compulsion. Therefore, agricultural policies designed to *mobilise* surpluses rather than *generating* them through increased agricultural productivity were bound to meet with failure in India. Yet, our policies revealed distinct preference for institutional measures which were inconsistent with the basic unalterables of the Indian situation. They betrayed a lack of awareness for raising agricultural productivity through increased investments and technical progress.

Public outlays for agriculture were reduced from one third of total outlays in the First Plan to 20 per cent and 23 per cent in the Second and Third Plans respectively. The pattern of these outlays also altered very much. Programmes designed to strengthen the capital base of agriculture, *e.g.*, irrigation projects and soil-conservation and other agricultural production programmes claimed as much as 90 per cent of agricultural outlays in the First Plan but their share was reduced to 60 per cent in the Second and Third Plans. The share of the rest of the programmes comprising mainly Community Development and Cooperation, on the other hand, increased correspondingly from 10 per cent in the First Plan to 40 per cent in the subsequent plans. Outlays for agriculture are a poorer indicator of achievement in real terms than the comparable outlays, say, for constructing a steel mill. Because of a large number of intermediaries involved between the planners and the farmers, the current or revenue expenditure has been higher and, consequently, the productivity of outlays has been lower. Moreover, administrative bottlenecks and uncertainties which are especially serious for agriculture, increased the time lag between the decision-making and the realisation of targets.

The planners were, however, confident that the institutional measures contemplated for agriculture would make good the gaps in public investments. These measures fall under two categories: land reforms, cooperative farming and heavier land taxation on the one hand, and Community Development Programme, on the other. The experience of the socialist countries was the source of inspiration for the former category of measures and of the technologically developed West, particularly the United States, for the latter. As argued below, both the categories of measures failed because in the one case they were not in keeping with the basic socio-political framework of India and in the other case the measures were based on erroneous preconceptions about the psychology of the Indian peasant.

Apart from the erroneous political assumption that land reforms as contemplated in the plans could in fact be carried through in India, the planners operated with several faulty assumptions on the factual plane as well. The basic idea behind such measures was that it would be impossible to effect a major breakthrough in agriculture without (a) ensuring security of tenure and 'fair' rents to tenants; and (b) breaking-up the large holdings and distributing the surplus lands to the small farmers and the landless who could later be induced to pool their lands for cooperative farming.

But the subsequent investigations, especially the National Sample Survey (NSS) and the Census of 1961, revealed that the area held by the tenants did not exceed 20 to 25 per cent of total cultivated area in the country, so much so, tenurial 'disincentives', if any, could not be a major retarding factor. Many of the landlords leasing out land are not absentees but cultivate a part of land themselves through hired labour and through intensifying modern inputs like fertilisers. Since the shortage of modern inputs limits their use beyond a point, landlords may find it more profitable to overcome managerial difficulties by leasing out their land to sharecroppers at this point instead of cultivating it through hired labour. In fact, the share-rent from tenanted land is found to correspond to the surplus from own cultivation after deducting costs including wages.

Thanks to the growth of population and the resulting subdivision of holdings, the area potentially available for redistribution after the imposition of ceilings was found to be negligible and the requirements of the uneconomic holdings and the landless too great to be capable of adjustment with any reasonable level of ceilings. Response for cooperative farming was negligible despite the several incentives provided for them in the Second Plan. Such of the societies as came into being were found to be non-genuine and were prompted either by the desire to evade ceiling legislation or to avail of the special privileges granted to them in the Plans. The measures to enhance land revenue failed because of wide-spread opposition. More recently, there has been a drive for the removal of land revenue at least on certain smaller holdings.

But the basic mistake of the planners was not that they were persuaded of the 'necessity' of such measures. One can still justify such measures in the interests of equity. And a more equitous distribution of land need not retard growth, if the necessary organisational measures are undertaken. Their basic mistake lay in equating what they considered to be a necessity with political feasibility. Given the socio-political framework within which planners and policymakers have to operate, there are naturally certain unalterable or relatively permanent elements in the situation. Therefore, the policy basket facing the planner should consist of measures which are *desirable as well as feasible or attainable*. The choice then exists between the attainable sets and it is here that cost-benefit considerations come into their own in the selection of efficient sets from among the feasible sets.

In regard to land reforms, the choice was made on the sole criterian of desirability to the exclusion of feasibility. The result was a systematic and successful evasion of tenancy as well as ceiling legislation through several ingenious methods. The most distressing consequence of these measures was that the administrative machinery reconciled itself to living with schemes which, it became sure, were not capable of implementation, and which, at the same time, it could not alter. Politicians reconciled themselves to making promises and passing legislations which, they were sure, would be evaded in practice. All this meant not merely a waste of scarce resources but the resulting frustration brought economic planning itself into disrepute.

The Community Development Programme was conceived not merely as an extension agency for transmitting new knowledge and supplying the modern inputs to the farmers but also as an agency for the education and the mobilisation of rural people towards a movement for their socio-economic betterment. The latter objective was prominent in the initial stages of the Programme and the impetus for this was provided by the impressive performance of refugee rehabilitation programmes around Delhi, which was obviously a special case involving people who were uprooted and were also known for their drive and hard work. Towards the close of the Second Plan it was realised that programmes for increasing agricultural productivity should claim as much as 80 per cent of resources of Community Development. movement. By this time almost the whole of rural area was covered by the Community Development agencies, without possessing, at the same time, either a knowledge of new techniques which the cultivators did not already know, or the requisite supplies of new inputs like fertilisers and pesticides.

The success of Agricultural Extension in the United States owed primarily to the new useful knowledge and inputs supplied to the farmers. The Village Level Worker in India failed to impress the farmer not merely because he was too young, urban-oriented and inexperienced in farming but also because he was not backed by adequate knowledge of new and profitable techniques and the requisite supplies of inputs. There has no doubt been an upsurge in rural areas for opening schools and hospitals and for constructing roads. But all this is traceable to the rise of agricultural classes and many of these programmes would have been carried through even without the Community Development movement.

While formulating agricultural programmes of the Third Plan, the failure of Community Development was attributed mainly to the lack of cooperation from the cultivators and the solution was sought in *Panchayati Raj* Institutions which ensured the supremacy of the popularly elected rural elite over the erstwhile Community Development administration. Indeed, the planners placed principal reliance on these institutions to carry out an ambitious programme of agricultural development envisaged in the Third Plan. The targets of the Third Plan in regard to food as well as cash crops were only slightly less than what was achieved during the decade 1951-61. The outlay on agricultural programmes provided in the Third Plan represented an increase of 92 per cent over the comparable outlay in the Second Plan. Although the programme could not be called over-ambitious in view of the immediate and pressing requirements of the economy, it is evident that relative to the Second Plan, the responsibility and the burden on the part of the administration as well as cultivators were almost doubled.

The sudden increase of responsibility in the Third Plan was a consequence of the neglect of agriculture in the Second Plan. But it was not realised that a belated action in the field of investments with a long gestation period can not be a substitute for the failure to space them adequately by undertaking some of these projects earlier in the Second Plan. The stagnation in agricultural output in the first three years of the Third Plan despite heavy investments should not have surprised the planners, because the realisation of targets in regard to irrigation and fertilisers is not a simple and instantaneous function of investments made. The roots of agricultural stagnation in the Third Plan lay really in the neglect of agriculture in the Second Plan.

Agricultural stagnation in the Third Plan forced the planners and the policymakers to seriously review the basic assumptions and methods underlying our agricultural policies. The rapidly growing population brought home the urgency of the problem and the increasing influence of agricultural classes on policymaking facilitated a more realistic approach towards agriculture. It is significant that the then Food Minister of Union Government* — the author of the new strategy of agricultural development based on high-yielding varieties — belongs to a progressive farm-family of a leading agricultural State. As a result of this re-thinking the emphasis has been shifting from the institutional reform of agriculture to the lifting of its

* Shri C. Subramaniam

technical base through the application of modern science. There is a growing confidence among the planners regarding the psychological preparedness of the Indian cultivator for modernising his vocation and an increasing realisation that he should be given full initiative in this process leaving to the public sphere the provision of basic infrastructure including the new knowledge through research and experimentation.

Agricultural Policy: Areas of Uncertainty

While the new strategy of agricultural development based on high-yielding varieties has caught the imagination of the planners and policymakers, there is, as yet, no concensus on certain key issues of agricultural policy for the future. What follows is an attempt to raise some of these questions and to indicate the line of action suggested by the experience of agricultural planning and performance in the past.

Consolidation and Strengthening of the New Strategy

The new strategy has no doubt raised hopes of a major agricultural breakthrough but it has not yet come to stay. The uncertainty regarding the new strategy consists not so much in the bottlenecks concerning the supplies of vital inputs like fertilisers as in the present state of ignorance concerning the whole complementary field *e.g.* pests and diseases, consequences for soil-structure *etc.* Owing to the pressing requirements of food, the natural tendency is to concentrate on those aspects which yield quick and substantial results to the relative neglect of aspects which require more effort and time but may, nevertheless, be crucial for sustaining the growth. The new strategy to be effective may require massive effort for research and extension concerning soil-testing, evolution of proper crop-rotations, and the identification of new diseases for finding remedies *etc.* In view of the soil and climatic specificities, the effort will have to be highly decentralised and diffused. The success of the new strategy depends on the extent to which these vital aspects are identified and the technical know-how strengthened and disseminated through the required investments and organisation.

Provision against the Uncertainties of New Strategy

The new strategy has strengthened the belief that substantial increases in output can be achieved with the limited water resources available through their intensive use in restricted areas. Therefore, the emphasis may shift from the creation of new irrigation potential to the provision of current inputs. It is still an open question whether the available water resources would be enough for achieving self-sufficiency on the agricultural front with the help of new strategy. In view of the tropical and sub-tropical character of Indian agriculture and the fact that as much as 90 per cent of cultivated area is directly or indirectly dependent upon monsoons which are highly erratic in India, the answer to this question is most likely to be in the negative. Notwithstanding the answer to this question, the problem still remains whether it would be prudent not to provide against the uncertainties of new strategy especially in the short-run. Tapping groundwater resources through quick-maturing irrigation projects provides a major buffer against such uncertainties, especially because the cultivators are familiar with the traditional techniques of irrigated farming. Therefore, minor irrigation schemes especially through rural electrification should continue to receive a high priority at least in the next decade.

Centre's Responsibility for Projects with a Long Gestation Period

With the progress of new strategy the problem of soil-conservation may assume increasing significance owing to the intensive use of land with imperfect technical knowledge about the soil-structure. Soil-conservation requires large investments, the results of which are not immediately visible. Major and medium irrigation projects and drainage schemes also belong to this category. The undertaking and execution of these projects requires imagination, sizeable effort and the preparedness to wait for the results of such a massive effort. All these have been a casualty with the State governments, who have been under continuous political pressure for multifarious schemes whose results are immediately visible. In fact, the serious shortfalls in respect of major and medium irrigation projects are attributable to such tendencies at the State level. Although agriculture is a State subject under the Constitution, past experience strongly suggests that such long-term projects should be undertaken and executed by the Central Government. This may call for the necessary adjustments — legal as well as financial — between the Centre and States.

Incentive Prices, Storage and Distribution

The recent stagnation in agricultural output and the sharp rise in prices have brought to forefront the problems concerning price and distribution policies. The situation called for a substantial procurement drive for foodgrains especially in surplus regions at prices lower than the abnormally high free-market prices and their equitable distribution among the deficit regions. This required not so much the expertise to clear the areas of ignorance and confusion in respect of policy as the willingness and capacity on the part of State and central governments to procure enough for the deficit regions. The performance in this respect has not been encouraging in the years of acute shortages. Consequently, the management of imported stocks represented a major effort in the sphere of pricing and distribution. However, with a major breakthrough in agricultural growth and the spatial concentration of surpluses entailed by the new strategy, policies concerning procurement, storage and distribution would assume increasing significance. Agricultural price policy may assume popular dimensions in view of the growing need to support farm prices and the favourable response for such measures. The increasing concentration of surpluses in the limited pockets of the surplus States may call for a carefully worked out and vigorous farm-price-support policy as the concentration of surpluses can depress the harvest prices even when the consumer-prices in other states rule very high. The storage programmes — whether on public or private account — would have to become procurement-oriented and undertaken at a much larger scale than envisaged so far, because the storage requirements would be much larger with a spatial concentration of surpluses than when the surpluses are diffused and find their way into pipelines. Further, the intensive application of modern inputs in the monsoon-based irrigated regions may increase the fluctuations in agricultural output which reinforces the need for building up buffer-stocks.

Programmes of Rural Industrialisation for Mobilising Agricultural Surpluses

The new strategy favours the irrigated regions and the rich farmers, owing to the complementarity between irrigation and new inputs, on the one hand, and the increased investment requirements, on the other. Therefore, the agricultural revolution, if successful, may create the problem of finding productive outlets for the surpluses of the prosperous sections as well as of providing more remunerative employment for the depressed sections. Past experience suggests that the mobilisation of agricultural surpluses through heavier land taxation or by turning the terms of trade against agriculture, should be ruled out in a democratic framework. Alternative strategies may have to be adopted for the voluntary channeling of agricultural surpluses and entrepreneurial talents into productive pursuits. A bold and imaginative programme of rural industrialisation may offer a solution. The basic bottlenecks for rural industrialisation is the lack of necessary infrastructure, *i.e.*, electrification, transport and communication facilities, banking institutions and training in skills. These are beyond the means of small entrepreneurs taking a start in rural areas and will have to be planned much in advance and provided for by the government.

Efficiency of Agricultural Administration

The exhaustion of autonomous factors in agricultural growth and the modernisation of agriculture with the active participation of the government will inevitably increase the dependence of agricultural sector on industry as well as the government. Indeed, agriculture has never been so much dependent on the external organisations for its supplies as it is now. To weather-uncertainty are now added the uncertainty of industrial performance and response of the administration. This may require a careful balancing of the relevant industrial targets in relation to agricultural requirements and the substitution of *Ad-Hocism* in government policies by carefully worked out measures with a view to ensuring greater durability of public policies. The increasing administrative responsibility for agriculture may call for a proper division of labour through greater reliance on the market-mechanism and the initiative of the farmer. This may involve the devolution of a greater part of the work concerning the production and distribution of inputs to the private sector. The administration can enhance its efficiency as well as the impact of its policies on agricultural growth by appropriating to itself only a limited number of crucial functions.

Published in N. Srinivasan (Ed.), Agricultural Administration in India, The Indian Institute of Public Administration, New Delhi, 1969.

3

Agricultural Growth and Stagnation in India

The object of this paper is two-fold:

(1) to analyse the trends in agricultural growth in the country as a whole and in different States during the period 1949-50 to 1961-62 with a view to bringing out the main ingredients of past experience; and

(2) to examine some of the current explanatory hypotheses relating to the relative stagnation of the agricultural sector. An attempt is made in this process to suggest the policy implications of the explanations that are found to be more relevant.

The All-India and State index numbers of agricultural production, area and productivity issued by the Ministry of Food and Agriculture fill a significant gap in available knowledge regarding the trends in growth. These index numbers are constructed after making due adjustments in the original data for changes in coverage and methods of estimation and are, therefore, comparable over time.

Two Seven-Year Periods

Separate growth rates have been estimated for two seven-year periods 1949-50 to 1955-56 and 1955-56 to 1961-62. These estimates reveal that the rate of foodgrains output had already slowed down from 4.4 per cent in the first seven-year period to 3.8 per cent in the later years, so that the growth rate of aggregate crop output remained more or less constant (4 per cent) despite an acceleration in the rate of growth of non-foodgrains from 3 per cent to 4.5 per cent. Another significant fact to note is that the growth of area under crops slowed down considerably from 2.8 per cent to 0.9 per cent between the two periods.

It would be interesting to know the relative contributions of area and productivity to the growth of output in different periods. If productivity per acre remained constant, output would grow at the same rate as the growth of area. Therefore, the growth of output attributable to the changes in productivity is given by the difference between the growth rates of output and area. The per cent contributions of area and productivity to the growth of output computed on this basis show that although for the 13 year period under study, area accounted for 48 per cent of growth in output, its contribution declined significantly from as much as 72 per cent to 23 per cent between the two periods. Thus, increases in productivity per acre accounted for as much as 77 per cent of output growth in the later period.

It would appear from this that the increase in area as a source of growth has almost exhausted itself in the Third Plan period and that further growth of output could be achieved mainly through the increases in productivity per acre. Although the growth of productivity accelerated between the two periods studied, it is clear that only a much higher rate of growth in productivity could have accelerated the growth of output in the Third Plan period. The Mid-Term Appraisal of the Third Plan revealed significant shortfalls in the supply of two main productivity increasing factors, *viz*., irrigation water and fertilisers. There was an element of spontaneity in growth so long as area was an important factor. Increase in productivity, on the other hand, requires input intensification involving large outlays — both public and private — as well as considerable organisational and administrative efforts. The difficulties associated with input-intensification in this new phase seem to lie at the root of the present stagnation in agricultural output.

Inter-State Disparities

The inter-State disparities in the growth of agricultural output are quite marked. The growth rate ranges from about 5 per cent per annum in the case of Punjab and Madras to just around one per cent for Assam, Orissa and West Bengal. That productivity per acre has become a predominant factor in growth is evident from the fact that all the seven States where growth of output is above the national average reveal higher growth rates in respect of productivity also. Except for Andhra Pradesh, the growth in productivity among the remaining eight States of slow growth was below the national average.

How would one explain these disparities in growth between different States? A State is a highly heterogeneous unit and may very often conceal significant differences in growth between its constituent units. Although the study of growth factors would be precise and more fruitful at a much disaggregated level, it should nevertheless be possible to discern factors which have a fair degree of validity over space. The hypotheses discussed in this connection fall broadly into two categories, *viz*., those which emphasise the institutional factors, more specifically the incentives associated with farm size and tenure conditions and those which emphasize material inputs – traditional as well as modern. It is true that the average size of farm is low in quite a number of States recording slow progress *viz*., Kerala, Uttar Pradesh, Assam, Orissa and West Bengal. But the net value of agricultural output per acre in these States is invariably higher than the All-India average. In fact, in the case of Kerala and West Bengal it is nearly twice the All-India average. Therefore, in a vast country like India where there are extreme regional variations in the soil-climate-crop complex, farm size is not a sufficient indication of income level. Moreover, Madras and Bihar where the average farm size is lower than the national average recorded satisfactory progress. Besides, Kerala with extremely low size of farm showed an impressive performance in regard to foodgrains while Uttar Pradesh was above the average in regard to non-foodgrains. It is hard to reconcile the resilience shown by these States with the 'disabilities' of size.

Since, smallness of size *as such* does not offer any technical hindrance to the application of modern inputs, the provision of credit to the smaller farms in a liberal way should meet the ends of growth. The enormous agricultural progress achieved by Japan despite the smallness of farm size should dispel many misconceptions in this regard. Although the large farms may be relatively quick in taking to improved techniques, progress may still be restricted owing to the managerial difficulties. The supply of mechanical power through small tractors and pump sets would promote the intensive use of land by lessening the managerial and supervisory difficulties. Such a technology may even increase the overall employment of labour by making possible the extension of intensive methods of cultivation such as irrigated farming to new areas. The structure of employment may alter but the volume of employment may not necessarily decline. The provision of credit to the small farmers in a liberal way and mechanical power to large farmers could well be the two main elements in a growth oriented size policy.

Tenancy

Estimates of cultivated area under pure tenancy are available for different states from 1961 Census. Considerable degree of precision attaches to these estimates as the information was elicited directly from the tenants themselves. Figures of area under mixed tenancy are not yet available but they may reasonably be assumed to vary between different States in the same direction as the area under pure tenancy. These figures do not reveal any precise relationship between the tenurial conditions and agricultural development. Although some States of slow overall growth, notably Kerala and West Bengal, account for very high ratios of cultivated area under pure tenancy there are significant exceptions notwithstanding the fact that it is difficult to reconcile this with the impressive performance of Kerala in regard to foodgrains. Pure tenancy is above the All India average for States of highest growth, *viz.*, Punjab and Madras. Similarly the ratios are quite low for many States revealing slow progress, *viz.*, Orissa, Uttar Pradesh, Rajasthan and Gujarat.

Fertility and Tenancy

The 1961 Census revealed that the proportion of tenanted holdings were particularly concentrated in the fertile coastal belts (see P.S. Sharma, "A Study of the Structural and Tenurial Aspects of Rural Economy in the light of 1961 Census," mimeographed). Among all the seven States where the proportion of area under pure tenancy is distinctly higher than the national average, the proportion of area irrigated is also well above the national average. In as many as five out of seven States where the area under tenancy is below the average, the proportion of irrigated area is also below the average. Further, the States where tenancy is significant, *viz.*, Kerala, West Bengal and Assam account for very high annual rainfall. It will also be seen that except for Punjab and Mysore, the remaining six States are characterized by high density of rural population. Also, the net value of agricultural output per acre for these States is distinctly higher than the national average.

Implications

These facts provide, in the first place, a rationale for the concentration of tenancy in fertile areas; land should be fertile enough to yield sizeable surplus to feed both the tenant and the landlord. Also, since the element of risk and uncertainty is considerably reduced because of the assured rainfall and irrigation, this would result in the high density of agricultural workers owing partly to the large-scale original settlements on fertile soils and also presumably to the higher rate of population growth resulting from the assured supply of food. All this would encourage tenancy because the rents would more or less be equated to the surplus that landlords would alternatively get by cultivating the lands themselves through hired labour.

The next important implication is that owing to the excessive demand for land, it becomes extremely difficult to obtain proper records of tenancy and to regulate rents through administrative action. Any administrative decree in this respect would most certainly drive tenancy underground with secret arrangements between the landlord and the tenant because of the real possibility of subsistence for both.

Lastly, from the standpoint of growth these facts should serve to dispel certain exaggerated notions regarding the discouragement of inputs among tenanted holdings. Since labour is abundant with hardly any opportunity cost, its application will add to the income of the tenant so long as the marginal product of labour is positive. The Farm Management Studies have revealed that the input of labour is highest for predominantly tenanted holdings and their output per acre is higher than for larger holdings operated through hired labour, as the input of labour among the latter can be extended only upto the point where the net marginal product is equal to the wage rate. There is no reason to believe that the fear of resumption of land by the landlord would discourage investments by the tenants in the input factors which are mobile such as bullocks and implements. In theory, the investments in permanent improvements on land would be discouraged but this becomes largely irrelevant when the tenanted holdings are situated in fertile coastal or irrigated areas. Again, the investments in modern inputs such as fertilisers, improved seeds *etc.* would, in principle, be discouraged if the tenants have to bear the full costs on them. But if landlords share such costs in the same proportion in which they realise rents from gross produce, then it would be profitable for both the tenants and landlords to invest in such factors until the marginal cost equals marginal revenue.

The practice of sharing inputs between landlords and tenants is not uncommon in this country. In fact, the so called "concealed share-cropping tenancy" is characterised by the system of sharing inputs. A recent study made by Wolf Ladejinsky in the Package districts where nearly 60 to 80 per cent of cultivated area is irrigated revealed that although about 50 per cent of cultivated area is under tenancy mainly through "off the record" arrangements, the expenditure on fertilisers is generally shared equally between the landlords and the tenants.

Japanese Experience

The Japanese experience in this respect has a particular relevance to Indian conditions. During the whole period of over half a century right from the Meiji Restoration to the close of Second World War, when the technological break-through in agriculture was effected mainly through the extensive use of fertilisers, improved seeds *etc.*, the area under tenancy continued to constitute nearly 50 per cent of cultivated area (See Takekazu Ogura Ed, "Agricultural Development in Modern Japan"). However, there is no gainsaying the fact that if somehow ownership rights can be conferred on the tenants in India such as had happened in Japan after the Second World War, agriculture would become more dynamic. But if past experience in India rules out this possibility for a reasonable period of time, then we need to have a more pragmatic and growth oriented tenure policy, the objective of which should be to promote input-intensification by insisting that the landlords share the expenditure on improved inputs at least to the extent of 50 per cent. Also credit should be extended to the tenants liberally in order to enable them to avail of these inputs.

Role of Irrigation

Among the material inputs, irrigation is known to bear the highest correlation with productivity. Unfortunately, the latest published statistics on irrigation in India do not extend beyond the year 1958-59. But some interesting information on this aspect is contained in the Food Ministry's "Growth Rates in Agriculture". According to this Bulletin, between 1952-53 and 1961-62, the increases that took place in the gross irrigated area in the first seven States of above average growth are much more than those in the States of slow progress. In Punjab and Madras gross irrigated area went up by 23 per cent and 33 per cent respectively; in Maharashtra it increased by 37 per cent; by 17 per cent in Madhya Pradesh; by 58 per cent in Mysore; and by about 15 per cent in Bihar. Among the States of slow growth, Rajasthan and Assam recorded some increase in the sown area without a corresponding increase in irrigated area leading to the negative growth of productivity. In Kerala, although there was some increase in the gross irrigated area, the per cent of net area irrigated to net area sown showed a marginal decline. Likewise in Gujarat although gross irrigated area increased appreciably, it failed to have significant impact on productivity because irrigated area constituted hardly 6 per cent of sown area. In Uttar Pradesh gross irrigated area showed a decline of 3 per cent. In Orissa and West Bengal too, there is no evidence of an increase in the proportion of area irrigated. Andhra Pradesh is the only state which despite an increase of 48 per cent in the gross irrigated area and the consequent increase in productivity, failed to show significant growth in overall production. This is explained by the fact that in the dry regions of the State where cash crops used to be grown extensively, net sown area declined by 1.5 million acres between 1956-57 and 1961-62 owing to the continuous failure of rains. On the whole it would appear that the growth in productivity and output in different States is found to be significantly associated with the increase in the proportion of area irrigated.

At present, the irrigated area in the country constitutes around 20 per cent of sown area. It has been estimated that about 50 per cent of cultivated area can be brought under irrigation (see Planning Commission, "Memorandum on the Fourth Five Year Plan"). It is evident from the Farm Management Studies that output per acre for irrigated crops is significantly higher than from dry ones ranging from 150 per cent to 400 per cent in different regions. Besides, irrigation increases output by making possible the cultivation of land more than once in the year. Assuming that, on an average, irrigation would raise output per acre by at least 200 per cent, aggregate output in the country would grow at the rate of 4 per cent per annum if two more acres are irrigated each year out of every hundred. According to a survey of minor irrigation made by the Programme Evaluation Organisation of the Planning Commission, over 60 per cent of cultivators in various parts of the country felt that the main problem faced by them in regard to irrigation was the construction of feed channels and the renovation of old tanks and wells. These rural works require community action on a big scale and it is here that the most difficult organisational problems are encountered. The breakdown of traditional leadership provided by the ex-landlords and the failure of Village Panchayats to fill in this vacuum has created a sizeable bottleneck in the sphere of community capital construction. Without planned direction and enforcements combined with necessary incentives it is doubtful whether the democratic institutions will, on their own initiative, apply themselves to these tasks. So far as irrigation is concerned it is of utmost importance to reduce the dependence on the usual administrative channels by entrusting this to the technical-cum-representative bodies at the district level constituted exclusively for this purpose with adequate resources as well as powers.

Technological Change

The role of science and technology in promoting agricultural growth in India is being increasingly realised. Theodore W Schultz in his recent book "Transformation of Traditional Agriculture" has advanced a rather persuasive hypothesis to explain the slow growth of agriculture in many of the underdeveloped countries. The basic elements of his hypothesis are: (1) the rate of return on the investment in traditional factors of production has declined to such a low level that there is no longer any incentive for cultivators in poor communities to save a larger proportion of their income to invest in such factors; (2) the only way to stimulate agricultural growth in such communities is to raise the investment opportunities by supplying new agricultural factors — both material and human - that embody improvements in technology and are therefore profitable for farmers to adopt; and (3) these two classes of investment are beyond the means of the individual cultivators and therefore require considerable public expenditures and the organisation of particular public activities to serve the agricultural sector.

As already mentioned, output can still be expanded profitably through the traditional irrigated farming. Even here, Schultz's hypothesis is relevant as the supply of water on an adequate scale is possible only through the modern means such as irrigation dams, tube-wells and pump sets which are beyond the means of individual cultivators. However, an important difference between these and the other modern factors is that once the water is provided, output can grow through the application of traditional skills at the farm level. It is heartening to note in this connection the crucial role assigned to science and technology in the future agricultural development of the country (see "Draft for the Fourth Plan"). However, without large scale public investments and community action at the village level for irrigation and soil conservation through afforestation and drainage schemes it is difficult to stimulate growth even through the use of modern factors of production as the profitability of the latter depends crucially on the availability of the former. Considerable public investments in the supply of modern inputs and improving human capabilities are nevertheless required right now for a number of reasons. For one thing, investments in research have a long gestation period and have, therefore, to be undertaken far in advance. Besides, farm children attending schools now will form effective labour force only after two decades or so and will remain in the labour force for several decades thereafter.

Considerable research is needed for evolving techniques suited to the Indian conditions. In a situation where there is considerable uncertainty and risk, and where the ruling interest rates are very high, the profitability of new techniques will have to be quite high for making them acceptable to the cultivators. It may be noted that the Rice Revolution in Japan and the Corn Revolution in the United States were the results of entirely new varieties of seeds suited to their local climate and distinctly fertiliser-consuming. Also, there was a technological break-through in the production of fertilisers leading to an appreciable reduction in their costs. The experiments conducted by the Indian Council of Agricultural Research have shown that under the existing price structure, the application of fertilisers was relatively more profitable for rice than wheat. This explains why the rice growing States of Andhra, Madras, Mysore and Kerala which had about 25 per cent of irrigated area of the country accounted for as much as 50 per cent of fertiliser consumption in the country in 1961 (See Planning Commission, "Third Five Year Plan"). On the other hand, Punjab which had about 9 per cent of the irrigated area of the country consumed only about 3 per cent of country's fertilisers.

Conclusion

These facts suggest the need for strengthening research both for evolving fertiliser-consuming seed varieties suited to different climates and for reducing the costs on fertiliser production. One of the main reasons why Community Development Programme failed to make an appreciable impact on agricultural productivity is that it was extended too soon to too many areas with too little of effective techniques at hand. The solution lies in evolving and promoting profitable technology and not merely in raising the status or salaries of Village Level Workers. All this implies that considerable resources should be allocated to scientific research. It is time the emphasis shifted from the role of agricultural surpluses in economic development to the role of science and industry in the development of agriculture. Agricultural surpluses would automatically emerge through the increase in agricultural productivity.

The Economic Weekly, February 27, 1965.

4

Problems of Agricultural Production in Kerala

The Problem

The acute food shortage in Kerala indicates neither the very low level of agricultural productivity nor the lack of dynamism among Kerala farmers. Curiously enough, the food shortage is the direct consequence of the considerable advances made in farming characterised by the extensive cultivation of commercial crops, which earn the valuable foreign exchange for the country besides bringing cash incomes to Kerala farmers. The available studies indicate that these cash crops are relatively more profitable than food crops in Kerala and that the experience and skills attained in the cultivation of such crops are a great asset for the country. It is, therefore, in the interests of Kerala and the country as a whole that the resources used *at present* are not diverted from cash to food crops. But the equilibrium can be maintained only if the country and more specifically the southern States discharge their *obligation* towards Kerala by ensuring sufficient and sustained supply of foodgrains. The experience in this respect has not been encouraging in the recent past. It is impossible to put up with recurring uncertainty in regard to such a vital item as food. The advantages of specialisation are bound to be outweighed, after a point, by the need to eliminate uncertainty in regard to food. Already, there is a clear tendency to divert resources from cash to food crops. During 1952-62 whereas agricultural production as a whole increased at the rate of 2.4 per cent per annum, the output of food crops increased at the rate of 4.1 per cent per annum while that of non-foodgrains increased only at about 1.8 per cent per annum. In the light of this experience, therefore, it is difficult to support the view on the other extreme that Kerala should specialise in the production of commercial crops by diverting more and more resources from food crops and meet its own food requirements through imports from other States.

The Perspective

The problem of diversion concerns the *existing* resources and as such represents a static approach to the problem. In the context of growth one is confronted with the dynamic problem of *increasing* the stock of resources with which to produce output. The problem would be less serious if instead of diverting the *existing* resources from non-food to food crops, more and more of newly created resources are used for food crops. The aim should be to reduce the magnitude of uncertainty in regard to food supply and not absolute self-sufficiency. The obligation of the rest of the country towards Kerala in regard to food is inevitable in view of the increasing need to earn the foreign exchange. The measures to increase food output in Kerala should, therefore, be viewed in the light of above considerations. The objective of applying more of new resources to food crops should be attained not by subsidising or pampering food crops' production and by making the production of non-food crops costly and more difficult. Such a method would be economically inefficient. Growth of food crops should be encouraged by creating conditions where investment in this direction becomes more profitable. This can be achieved by rational investment and tenure policy. In this context, the major problems of agricultural development facing Kerala are discussed below.

The Basic Measures

(a) Soil Conservation: Soil erosion is a serious problem in Kerala owing to its vulnerable topography and other factors. According to the National Council of Applied Economic Research, about one-third of cultivated area is affected by soil-erosion. Soil-conservation is an indispensable condition for raising the profitability of the application of fertilisers, improved seeds etc. The task of soil-conservation may evade attention because (i) it is beyond the means of individual cultivators and requires considerable organisational efforts at the local level involving community action with Government's participation; (ii) the costs are sizeable; (iii) it requires sustained work with long gestation period *i.e.,* the results are not immediately visible. But this problem can be neglected only at heavy cost because poor soils depress the profitability of all the complementary inputs and since soil-erosion is a continuous process, any delay in this respect adds to the cost of its removal. It is, therefore, of utmost importance to overcome this vicious circle by undertaking a bold programme of soil-conservation before the costs become prohibitive.

(b) Irrigation: A high level of rainfall is not an unmixed blessing. Since artificial irrigation in Kerala is required generally for second or third crop, any large scale irrigation system becomes uneconomical. Even minor irrigation is not as economical in Kerala as in areas of relatively low rainfall. However, minor irrigation sources are quite important in Kerala. Besides, high rainfall lends a kind of stability to minor irrigation in regard to water-supply in this State which is not possible in the States of low rainfall. In order to encourage cultivators to take to irrigation through minor works on their own, it is essential to make irrigation cheap. Supply of electric power and electric motors provides the cheapest and the most convenient means of irrigation for the cultivator. In regard to the generation of cheap power, Kerala is most fortunately placed. Here again, an imaginative and bold programme of supply of power for

irrigation would go a long way in stimulating and sustaining agricultural growth.

(c) Improved Techniques: Although there has been some progress in the use of modern inputs such as fertilisers and improved seeds in Kerala, the performance is far from satisfactory in regard to food crops and there exists considerable scope for the adoption of improved techniques. Indeed, Kerala with its limited land supply, abundance of labour and high rainfall presents a case of the near exhaustion of traditional input factors. The case for adoption of improved techniques as the only possible method of increasing productivity in India is nowhere as urgent as in Kerala. At the end of Second Five Year Plan Kerala accounted for about 3.6 per cent of irrigated area in the country as compared to 6 per cent in the neighboring rice-growing State of Madras. But the share of Kerala in the consumption of fertilisers was much lower on a proportionate basis. It consumed a little less than 3 per cent of Ammonium Sulphate and a little less than 2 per cent of Super-Phosphate consumed in the country. The share of Madras, on the other hand, was 12 per cent in the case of Ammonium Sulphate and around 15 per cent in the case of Super-Phosphate (see Third Five Year Plan). This low consumption may be attributed partly to the high price of fertilisers. This problem is common to the whole country and the reduction of fertiliser price, though very essential, is beyond Kerala's responsibility and the benefits in this respect will have to come from the national endeavor. Partly, this lower consumption may be attributed to the lower return on the relatively infertile and eroded soils. The experience in the United States as well as Japan shows that fertiliser application becomes profitable when it is combined with the new and improved seed varieties that are fertiliser-consuming and pest-resisting. In regard to the use of improved seeds for food crops the position in Kerala is not encouraging. Towards the end of the Second Plan the area under improved seeds (food crops) in Kerala constituted less than 1 per cent of that in the country as against 13 per cent in Madras. This may be explained in part by the fact that in Kerala greater

attention is devoted to commercial crops. But there is no gainsaying the fact that the improvements in the food crops' productivity in Kerala depend crucially on the application of improved inputs. So far as rice culture is concerned, Kerala can become a major centre for research, experimentation and extension in the improved technology. The pressing food requirements in Kerala suggest not only the urgency of sizeable investment in this respect but also give promise of a phenomenal response to such efforts. The experience in this respect can prove to be useful for the rest of the country.

(d) Tenure Policy: According to 1961 Census, as much as 35 per cent of cultivated area in Kerala is under pure tenancy, as compared to 4 per cent for All India. Since this excludes the tenanted area which is cultivated jointly with the owned area, it is not unlikely that nearly 50 per cent of cultivated area is under tenancy. This may account for the insufficient application of fertilisers and improved seeds in as much as the costs are borne by the tenants and benefits reaped by landlords in an equal measure. This problem can be solved if landlords are required to bear the costs on such improved inputs in the same proportion in which they claim rent from the gross output. There is no reason why landlords should not agree to share the costs because so long as rent bears a fixed proportion to gross output, they would benefit from the increments in output. So far, the tenurial policy in India has been oriented to equity goals with its insistence on the security of tenure and fixation of fair rents etc. It is seldom realised that under conditions of excessive pressure on land, it is extremely difficult to implement such a policy. Not only that the equity goals are not achieved but in the process the growth objectives are also defeated. It is, therefore, of utmost importance to take a realistic view of the situation and pursue a growth-oriented tenure policy, the objective of which should be to promote input intensification by ensuring that landlords share at least 50 per cent of costs on improved inputs. The tenants should be provided the necessary credit facilities to

meet their share of costs. The Japanese experience in this respect has a particular relevance to Kerala conditions. During the whole period of half a century right from the Meiji Restoration to the close of Second World War, when the technological break-through in agriculture was effected mainly through the extensive use of fertilisers, improved seeds etc., the area under tenancy continued to constitute nearly 50 per cent of cultivated area (see Takekazu Ogura Ed., "Agricultural Development in Modern Japan"). In India, no place is in greater need as well as better suited than Kerala for experimenting a pragmatic and growth-oriented tenure policy.

Conclusion

It has been argued above that the increases in food output in Kerala should be achieved not by diverting the *existing* resources from commercial to food crops but by allocating more and more of *new* resources to food crops. Such an allocation pattern should be ensured not by subsidies and controls but by creating conditions where growing of food crops becomes more profitable. This can be achieved by making soils more productive through soil-conservation and irrigation; by supplying low-cost improved inputs through research and experimentation; and by providing incentives to the tenants through persuading landlords to share at least half the expenses on improved inputs. It is clear that none of the measures discussed above is likely to be an easy undertaking offering quick results. But this is as it should be in a situation where the productivity from the traditional sources has almost reached a saturation point. At the same time, the potentialities from improved techniques should be quite promising. What is required is a bold and imaginative policy backed by sustained efforts and that which does not fail for want of immediate successes.

Seminar on Agricultural Production, Kerala, Trivandrum, 1965.

5

Economic Consequences of the New Rice Technology: General Comments

The several papers presented at this Conference provide a wealth of data based on experience, particularly in the Philippines. One is impressed by the enormous amount of work of the IRRI economists. One wonders how such a mass of work could be accomplished. The answer seems to be the availability of useful data at the IRRI as well as the comparative freedom that the undoubtedly competent IRRI economists enjoy from the distractions of the rest of the world! The IRRI studies, which throw light on various aspects of the consequences of the new rice technology, are bound to interest readers for a long time to come and should stimulate further research and thinking on a subject of considerable importance to the peoples of South and Southeast Asia.

I discuss the major points emerging from conference papers and discussions and identify what I consider as major gaps in the research done. Such an approach, I hope, will provide a perspective for future lines of work.

One is struck by the enormous difference between the economically attainable potential yields of rice and its actual yields. Herdt and Wickham show the gap as almost three-fourths of the economically attainable potential. They feel that about three-fourths of the gap is realisable. Their analysis suggests three factors as equally important in realising the potential: irrigation and flood control, factors such as credit and interest rates that bear on risk, and research to evolve technologies that can overcome environmental constraints. The Herdt and Wickham analysis suggests the need for

- investment in infrastructure such as the provision of new irrigation - particularly for the dry-season crop - and flood control,
- reform and development of institutions providing credit and extension services, and
- research to develop technologies to fit local conditions and unfavourable environments.

All these pose equally important challenges for governments, scientists, and economists.

Investment in Infrastructure

The importance of investments in infrastructure such as irrigation and flood control is the most significant finding of these studies. A number of papers bring that out. Wickham and Barker note the high complementarity between irrigation and modern varieties (MV). They show that the yield performance is substantially better than that of traditional varieties (TV) at higher nitrogen levels with irrigation, particularly irrigation with high dependability. The shift from TV to MV makes the benefits attributable to irrigation 50 per cent higher where there is efficient use of fertiliser and water.

Kikuchi and Hayami show that the profitability of investment in irrigation has increased recently and has induced government investment in irrigation systems. That has happened despite the increase in cost, or decreasing profitability trends, resulting from the completion of less costly projects. The rise in profitability of investment in irrigation is explained by the steep decline in per capita arable area, the increase in the marginal cost of opening new land, and the diffusion of the seed-fertiliser technology since the late 1960's. To these is added the fact that the available irrigated land is scarce and has been used intensively with the application of MV and fertilisers, which increases the rate of return on further investment in irrigation.

The increase in the international price of rice has also raised the profitability of investment in irrigation as a means of attaining food self-sufficiency. That is brought out by Barker, Bennagen, and Hayami.

David and Barker show that the demand for fertilisers is highly sensitive to price changes particularly in countries with high levels of fertiliser use, but Desai points out that the price elasticity may have been overestimated owing to the non-inclusion of the diffusion factor representing the movement along the production function. Nevertheless, a significant response of fertiliser demand to changes in price is quite likely when a complementary input like irrigation is limited and the application of MV and fertilisers is concentrated in limited areas where the necessary institutional infrastructure is available and where, therefore, the product curves flatten out.

On the other hand, the yield response to price of rice or to the rice-fertiliser price ratio has increased as a result of the introduction of MV and the development of irrigation systems, which has the effect of making the application of fertilisers and related inputs more responsive. This result implies that the development of irrigation creates the necessary condition for the effective operation of policy variables regarding the input supply and prices. This is clear from the study by Sison, Somsak, and Hayami on structural changes in rice supply relations.

The rise in the rate of return to investment in irrigation, the high sensitivity of fertiliser consumption to prices, and the rise in the output response to prices derive from the complementarity between irrigation on one hand, and the MV and fertilisers on the other. Thus a high sensitivity of fertiliser consumption to price may reflect either the insufficiency of investment in irrigation, or a misallocation of fertilisers from the social point of view (i.e. concentration in limited areas), or both. In either case, the implication would be to increase investments in irrigation and other related infrastructure, and to improve the allocation of resources through institutional changes. Mere reduction in the relative price of fertilisers so as to increase their consumption will only accentuate the misallocation of resources by further depressing the production per unit of fertiliser.

Constraints

The papers also suggest the need to effect a change in the research focus toward overcoming environmental constraints. As mentioned earlier, the analysis of the constraints to the realisation of the potential of new rice technology suggests that environmental constraints are a major factor and account for at least one-third of the gap between the economically attainable potential and actual yields.

More research is therefore needed to evolve flood-resistant and pest-resistant rice varieties and technologies, and cropping patterns that maximise income and employment per hectare per year. A visit to IRRI experiments on the first day of the conference was an eye-opener in this respect. The focus on research was on methods to overcome the environmental as well as capital constraints. Such research is bound to benefit regions poorly endowed in respect of climate and physical factors as well as capital resources. I venture the opinion that the biological scientists are now ahead of the economists in responding to the needs of the unfavourably placed regions and classes.

It has been argued at this conference that the payoffs to investments in such research may be less attractive than to efforts towards the development of new technology for favourably placed regions. This hypothesis needs to be tested.

It is also suggested that investments in physical infrastructure, wherever it can be built up, would be more rewarding than investment in research to overcome environmental constraints arising largely from an inadequate infrastructure. All these issues need further examination.

There is no evidence to date to rule out as unprofitable the investments in research to overcome environmental constraints. The available evidence suggests that it has been profitable to evolve and adapt technologies to suit different environments and resource situations. The study on rice research by Evenson and Flores reveals extraordinarily high returns from investment and indicate that too little investment has been undertaken in the past, particularly by national governments.

Employment Impact

The impact of the new rice technology on employment and distribution of income has not been brought out clearly in the conference papers. The only exception is the analysis by Hayami and Herdt on the favourable distributional effects via lowering the output price. The new technology, insofar as it brings down the relative price of food, improves the real income of the urban and rural wage earners and small farmers who spend most of their income on food, whereas the real incomes of the rich farmers come down because of the decline in food prices. However, Hayami and Herdt point out that a real danger would arise if new technology is monopolised by a small number of large producers and does not cause a significant shift in the aggregate supply schedule. In such a case, the large farmers could capture the whole gain of technological progress by increasing output without a resultant decline in prices.

I have pointed out elsewhere (Rao, 1975) that this is exactly what happened in India because of the deficiency of public investment in infrastructure. The indivisible factor, such as irrigation, as Hayami and Herdt point out, may be too large for purchase by individual small farmers.

The relationship between farm size and employment per hectare was not analysed systematically. Also there is no investigation into the patterns of regional migration of labour under the impact of new technology, and the effect of technology on the aggregate employment in the developed and depressed regions together.

The discussion on the employment effects of new technology, particularly tractorisation, has revealed certain methodological deficiencies. The typical approach to this question is to compare farms adopting new technology with those not adopting and then examine the differences in employment. Correction is, of course, made for the differences in employment arising from variation in the complementary inputs. It should not be surprising that many of the studies reveal that the impact of tractorisation on employment is positive. Such an approach evades the basic question of capital:labour ratio or the choice of techniques.

In the Punjab, for instance, the output of agricultural commodities as a whole has been growing by about 8 per cent a year. In the absence of mechanisation, such an output would have required a labour force with a growth rate of about 6 per cent a year, on the assumption that the labour coefficient is .75. However, the labour force within the Punjab may not have grown by more than 2 per cent a year. The gap of about 4 per cent has been filled partly by the migration of labour from regions such as eastern Uttar Pradesh and partly by mechanisation. Wage rates in the Punjab have been rising despite in-migration and the increase in employment because of the economic and psychological costs associated with migration both for the in-migrant labour as well as for the farmers employing them. Mechanisation on a large scale has, therefore, proved profitable. The fact that mechanisation has been associated with significant increase in employment in the Punjab should not detract from the fact that the aggregate employment for the agricultural sector as a whole would have been greater if the capital:labour ratio were smaller. That could have been made possible with a more balanced regional development of agriculture, through public infrastructural investments in the lagging regions, and through appropriate input price policies. My study (Rao, 1975) shows that the labour coefficient was highest for small-scale cultivation; it was followed in descending order by that for cropping intensity, irrigation, and MV with fertilisers. Tractorisation had the lowest labour coefficient, which did not appear to be significantly different from zero.

The conference papers discussed were not concerned with such institutional factors bearing on employment and income distribution as the supply of credit, interest rates, tenancy, and farm size. Labour absorption under different techniques, such as extension of area, irrigation, fertilisers, cropping intensity, and tractorisation, did not receive adequate attention either.

Ingredients of Underdevelopment

The conference papers and the experience with the new rice technology clearly underscore the importance of three common ingredients of underdevelopment - low level of investment, traditional technology, and outmoded institutions.

The growth models of the 1950's and early 1960's emphasised investment with institutional changes in the agricultural sector as a strategy for growth in an underdeveloped economy.

Experience has shown a large degree of complementarity between growth and distribution in the developing economies. Despite the introduction of new technology, agricultural growth has been slow because of a narrow base caused by insufficient investment, and by outmoded institutions. Stepping up public investment and institutional changes, including those in administration and organisation, would not only promote growth but improve the distribution of income as between different regions and classes of farmers.

In India, for example, public investment in the exploitation of surface water for irrigation would reduce regional disparities in income, and benefit all classes of farmers in proportion to the area they hold. Public investment in the exploitation of groundwater in the Indo-Gangetic plains would benefit millions of small and marginal farmers who cannot exploit the water on their own because of inadequate resources.

Failure of public investment in the infrastructure, insofar as it results in slow growth, will cause a rise in output prices. Such a rise in prices may stimulate private investment in factors including irrigation. However, such a strategy would lead to poor results in terms of both growth and distribution. Growth is bound to be slow because the scope as well as capacity for private investment in agriculture is limited. Only a few affluent farmers can undertake such investment. The experience in India indicates that distribution of public sources of irrigation is less unequal than that of private sources.

Because the conference has not faced squarely the question of factor-use bias in technology, the distributional question has been neglected. Hayami and Ruttan (1971) pointed out how agricultural techniques tend to be adapted to the factor endowments and factor prices of each country, and emphasised the need for land-augmenting techniques for the countries of South and Southeast Asia. I have argued (Rao, 1976) that techniques that increase output per hectare with the intensive application of fertilisers along with new seed are not necessarily best suited to the factor endowments of those economies. The

South and Southeast Asian economies need land-augmenting techniques that maximise the use of their abundant factor - labour - and economise on the use of scarce resources - capital. That may involve the use of techniques such as irrigation and multiple cropping to maximise output per hectare per year by maximising employment per hectare per year. The focus so far has been to use new seeds and fertilisers to maximise output per hectare per crop.

Reshaping of Institutions

Apart from the evolution of techniques, the reshaping of the existing institutions affecting land use and credit may be essential for meeting the objectives of growth and improved distribution. Credit institutions, administrative organisation for supply of inputs, farm tenancy, and farm size impinging on land-use patterns are important in this context. I agree with Ruttan that institutional changes to make the most effective use of the labour resource, which is an abundant factor in these economies, should receive the greatest attention in the next decade. This objective can be achieved only through institutional changes involving greater supply of capital resources to the small farmers (who have more labour resources) and by evolving techniques that maximise the employment of labour per hectare per year. Farm size must be appropriate in terms of maximum labour absorption. Mechanisation also belongs to this area. Ruttan's proposition has to be translated concretely into such policies and strategies.

It has been argued that IRRI is not the appropriate place for economic research bearing on such institutional questions. I believe, however, that it is precisely in a place like IRRI that such research becomes essential, because of the need for interaction between the biological scientists and economists for evolving technologies suited not only to specific environments but also to specific resource situations. IRRI scientists appear to be already on the way to doing good work bearing on the distributional aspects. Such work by the economists should be strengthened further through appropriate economic analysis of the interaction between the factor-use bias of technology and the institutional factors that bear on the resources and resource prices. Incidentally, the authors quoting the Indian examples seem to ignore the enormous amount of empirical evidence on the distributional aspects of new seeds in Indian agriculture.

Economists have done several studies recently to show how farmers of the developing economies are *rational* in the sense that they respond to the changes in the relative prices of crops by reallocating their resources. Kikuchi and Hayami show how even governments are *rational* insofar as their investment decisions respond to changes in prices. They, however, point out that this apparent *rationality* may be traceable to the basic irrationality represented by the deficiency in the allocation of government resources earlier, which has made public investment in irrigation highly profitable. It is clear now that an apparently *rational* behaviour, whether by government or peasants, in the cob-web fashion, is not inherently rational. Unfortunately, experience in the last two decades shows that economists too tend to be 'rational' like peasants and governments, by responding to changing prices or current crises.

Insensitivity of economic analysis to changing needs would render economists irrelevant, but it must be recognised that the given situations are often the result of lack of perception, imagination, and forethought in the past, whether in scientific analysis or in public decision making. Economists as social scientists can be more perceptive and imaginative than the short-run income-maximising farmers or the security-conscious public decision makers.

References

Barker, R., E. Bennagen, and Y. Hayami, (1978). "New Rice Technology and Policy Alternatives for Food Self-sufficiency." Pages 337-361 in *International Rice Research Institute*, Economic consequences of new rice technology. Los Banos, Philippines.

David, C.C., and R. Barker, (1978). "Modern Rice Varieties and Fertiliser Consumption." Pages 175-211 in *International Rice Research Institute*. Economic consequences of new rice technology. Los Banos, Philippines.

Evenson, R., and P.M. Flores, (1978). "Social Returns to Rice Research." Pages 243-265 in *International Rice Research Institute*. Economic consequences of new rice technology. Los Banos, Philippines.

Hayami, Y., and R.W. Herdt, (1978). "Market Price Effects of New Rice Technology on Income Distribution." Pages 283-302 in *International Rice Research Institute*. Economic consequences of new rice technology. Los Banos, Philippines.

Hayami, Y., and V.W. Ruttan, (1971). *Agricultural Development: An International Perspective*. Johns Hopkins. 367 p.

Herdt, R.W., and T.H. Wickham, (1978). "Exploring the Gap between Potential and Actual Rice Yields: The Philippine Case." Pages 3-24 in *International Rice Research Institute*. Economic consequences of new rice technology. Los Banos, Philippines.

Kikuchi, M., and Y. Hayami, (1978). "New Rice Technology and National Irrigation Development Policy." Pages 315-332 in *International Rice Research Institute*. Economic consequences of new rice technology. Los Banos, Philippines.

Rao, C.H. Hanumantha, (1975). *Technological Change and the Distribution of Gains in Indian Agriculture*, Macmilan, New Delhi, 249 p.

——— (1976). "Factor Endowments, Technology and Farm Employment: Comparison of East Uttar Pradesh with West Uttar Pradesh and Punjab." *Economic and Political Weekly*, Review Agric. (September 25) 11(39):A-117, A-119, A-123.

Ruttan, V.W., (1978). "New Rice Technology and Agricultural Development Policy." Pages 367-381 in *International Rice Research Institute*, Economic consequences of new rice technology, Los Banos, Philippines.

Sison, J.F., S. Prakongtanapan, and Y. Hayami, (1978). "Structural Changes in Rice Supply Relations: Philippines and Thailand." Pages 31-48 in *International Rice Research Institute*. Economic consequences of new rice technology. Los Banos, Philippines.

Wickham, T.H., R. Barker, and M.W. Rosegrant, (1978). "Complementarities Among Irrigation, Fertiliser and Modern Rice Varieties." Pages 221-232 in *International Rice Research Institute*. Economic consequences of new rice technology. Los Banos, Philippines.

Economic Consequences of the New Rice Technology, International Rice Research Institute, Los Banos, Philippines, 1978.

6

A New Look at Agriculture in the North-West: How to Overcome Limits to Growth

The experience of agricultural development during the last decade and a half in the country shows clearly that the performance of Punjab and Haryana has been about the best. The growth of farm output in this region has been far above the national average.

As much as 90 per cent of the geographical area in Punjab and Haryana has low rainfall (up to 750 mm.), but the proportion of cultivated area irrigated is highest in the country. The combination of abundant sunshine and controlled irrigation has facilitated the adoption of high-yielding technology for agriculture. The institutional infrastructure represented by agricultural credit institutions and markets is also highly developed in Punjab and Haryana. Above all, the human factor for development, as reflected in the initiative and enterprise of the people in economic endeavour, has played no mean part in achieving impressive progress.

Because of the rapid growth of agriculture, Punjab and Haryana have reached the highest level of per capita income among all the states in the country. The proportion of income from agriculture to total income is also among the highest (around 58 per cent). This agriculture-based growth explains why the proportion of the people below the poverty line is the lowest in these two states (about 27 per cent for Punjab and 37 per cent for Haryana) than in any other.

Although agriculture in this region is more capital-intensive than in other parts of the country, the overall capital-output ratio in this region — i.e. output produced per unit of capital invested — is also perhaps more favourable than elsewhere because of the predominant contribution of agriculture, which, in general, has a lower capital-output ratio than the non-agricultural sector.

Has Agriculture Development Reached its Limits in North-Western Region?

The agricultural development in this region has proved to be a major source of employment for the poor migrating from the less-developed regions in eastern India, although employment in terms of man-days per acre and per unit of output is still low in this region compared with other areas in the country similarly served with irrigation facilities. The out-migration of skills, enterprise and capital from this region has also been fairly significant both within and outside the country.

The inevitable question that arises is whether the limits to agricultural development have been reached in the plains of the north-western region. The recent spurt in the growth of rice can be explained by favourable climatic factors as well as by the limits to growth reached in the case of wheat. Within the constraints of existing technology, a saturation point appears to have been reached in respect of investment in fertilisers and other inputs for wheat. With the given technology, therefore, further growth of wheat output in this region can be achieved only by increasing unit costs which cannot be met without raising output prices.

Rice, on the other hand, seems to have further potential for growth even with the existing technology and prices which are fixed after taking into account production costs in the high-cost regions like Tamil Nadu and Andhra Pradesh. However, it is reasonable to expect that the limits to growth in the case of rice too may be reached soon, as the potential for extending irrigation gets exhausted.

It is highly probable, therefore, that it would be difficult to sustain a high rate of growth of agriculture in the north-west with relative stability in prices. A high-growth in agricultural output can come only through increases in unit costs supported by higher output prices.

In a situation where prices and wages are rising, one should expect the farmers to try to increase their profits by reducing costs through the substitution of capital for labour. This is a real possibility, particularly because this region, for various reasons, has high investible surpluses. Such a move towards higher capital-intensity through greater mechanisation — particularly of harvest operations which account for the highest wage rates — may hit the poorer workers who are attracted to Punjab and Haryana from the less-developed regions.

If this situation is to be avoided, capital and enterprise from the north-west will have to find profitable outlets elsewhere consistent with the national objective of stepping-up agricultural output through balanced regional development, along with adequate generation of employment and purchasing power for the poor.

Rural-to-rural migration of labour has been the most important form of migration in this country and, Punjab and Haryana have had an impressive record in this respect. Ironically, the green revolution has enabled many a farmer in this region to look for even better prospects overseas. Such a migration represents the drain of scarce resources like skills and enterprise for agriculture, and may, therefore, result in a significant loss to the country, despite the high value attached to the remittances from abroad.

Instead, we need to encourage migration to other parts of the country by providing the necessary incentives in the sphere of agriculture. It is possible that such migration of skills, enterprise and capital to the less-developed regions with a significant potential for growth would yield greater benefits in terms of increase in overall productivity than it would if these resources are concentrated in the few developed areas.

The gains will flow not only from the increases in the productivity of resources transferred but also and more importantly from their demonstration effect on the local farmers as well as the extension workers who have the responsibility to make use of the funds and farm inputs allocated for the less-developed regions.

Such a planned migration, insofar as it provides outlets for the surpluses and skills from Punjab and Haryana, would also help restrict "conspicuous consumption" in the two states. Indeed, the indirect employment created in the developed areas, following technological changes, seems to be both a result of and contributory to such consumption. In the long run, it is bound to increase regional disparities.

The resistance to the in-migration of capital and skills is likely to be insignificant if it is regulated, for example, by putting a ceiling on the proportion of land that migrants can hold. The "sons-of-the-soil" movements in India are mainly directed against the provision of employment in the public sector to the so-called outsiders and are not so much against private investors from outside the region who, in fact, contribute to increasing employment opportunities for the local people.

The other outlet is industrial development, including rural industries in this region itself. Historically, for various reasons including uncertainties in the border region, not many major industrial enterprises have come up in the area. However, the performance and potential of the north-west in the sphere of small-scale industry, particularly ancillaries, is well-known.

It is also recognised that social development in the region in terms of the growth of education, especially female literacy, and improvements in health and sanitation, have not been commensurate with the growth in material output. The potential for social development may need to be better exploited in the next decade.

Role of State

The scope for rural industries also needs to be explored. The transition from agriculture to industry is not easy in a society characterised by "peasant culture". Such a transformation requires reorientation of skills as well as re-adjustment of entrepreneurial motivations and attitudes.

In a situation like this, the role of the state in providing the necessary physical and institutional infrastructure and leadership in the organisation and running of corporate bodies involving farmer interests becomes decisive. Notwithstanding the well-known limitations of the bureaucracy, its contribution to development in this region has certainly been greater than in most other parts of the country. The participation of the state in rural industrialisation can contribute considerably to the mobilisation and harnessing of rural surpluses in Punjab and Haryana for productive purposes.

The Times of India, July 22, 1981.

7

Fluctuations in Agricultural Growth: An Analysis of Unstable Increase in Productivity

It is difficult to conceive of a neutral relationship between agricultural growth and the amplitude of weather-induced fluctuations in yields. Nor should there be any unique relationship between the two. Yield variability may depend on the pattern of investment or the methods by which growth is brought about. For instance, irrigation water from perennial sources (*e.g.*, major river valley projects and tubewells) and use of pesticides may counteract the vagaries of weather and if increases in agricultural yields are brought about mainly through investments in such inputs, then growth may be associated with a decline in the amplitude of fluctuations. Provision of irrigation water through tanks or ponds and rain-fed wells, on the other hand, may accentuate the fluctuations by increasing the dependence on weather. While these sources cannot provide an adequate cushion in the event of failure of rains, they may, nevertheless, push up yields in a year of good rainfall by making possible the storage of water. Thus, growth in productivity per acre through an increase in the proportion of area irrigated by tanks and rain-fed wells may have a built-in tendency to increase yield-instability.

There are inputs like labour, fertilisers and improved seeds which, if used under conditions of assured irrigation, may promote growth with stability but, if used under conditions of uncertain rainfall or in areas irrigated by tanks and wells, may increase the range of fluctuations in growth. Owing to the complementarity between these inputs and water, output may be pushed up significantly in a year of good rainfall while, in the event of failure of rains, they are unlikely to provide a significant cushion, thus, widening the gap in yields between good and bad years.

Higher Productivity

The present study is concerned with the analysis of fluctuations in production as well as in yields or productivity per acre. A study of output fluctuations alone would not give a clue to the built-in variability of yields per acre since the relative importance of productivity per acre has been increasing recently. Over the period under study, productivity per acre has fluctuated much more than sown area (Table 7.1) and output has grown mainly through increase in productivity per acre. Between 1956-57 and 1961-62, agricultural output (all crops) increased at the rate of 3.4 per cent per annum whereas area under cultivation increased at 0.8 per cent per annum (Table 7.1). If productivity had remained constant, output would have grown at the same rate as sown area, *i.e.,* by 0.8 per cent. The balance of output growth, *i.e.,* 2.6 per cent per annum or about 76 per cent resulted from increases in productivity.

This study is concerned with fluctuations in overall production and productivity indexes rather than in individual crops. Productivity variations in individual crops may derive partly from changes in relative prices. Second, weather is known to affect the yields of different crops differently, in respect of direction as well as magnitude. Productivity of individual crops, thus, tends to fluctuate much more than that of crops in the aggregate (Table 7.2).

Table 7.1

Growth and Fluctuations in Area, Production and Productivity

(All Crops, All-India)

	1952-53 to 1957-58	*1956-57 to 1961-62*	*1952-53 to 1961-62*
Area Under Crops			
Growth Rate (linear)	2.8	0.8	1.3
Coefficient of Variation	1.4	1.0	2.1
Production (All crops)			
Growth Rate (linear)	3.9	3.4	3.2
Coefficient of Variation	4.0	4.7	4.4
Productivity per acre			
Growth Rate (linear)	1.0	2.6	1.8
Coefficient of Variation	2.7	3.3	3.4

Source: Government of India, Ministry of Food and Agriculture, "Growth Rates in Agriculture", 1964.

Table 7.2

Coefficient of Variation in Productivity Per Acre of Individual Crops

(1952-53 to 1961-62)

Crop	*Coefficient of Variation*
Rice	6.3
Jowar	8.7
Maize	6.2
Wheat	6.5
Gram	9.1
Oilseeds	6.6
Cotton	9.4
Jute	5.0
Sugarcane	5.4
All Crops	**3.4**

Source: Government of India, Ministry of Food and Agriculture, "Growth Rates in Agriculture", 1964.

Fluctuations in productivity are measured by the coefficient of variation. Since coefficients of variation would be different for the series showing differing trend movements, it is necessary to make correction for such trend movements. This is done by measuring the deviations in productivity for each year from the respective trend values instead of from the mean value. The overall mean for the period is then used for computing the coefficient of variation.[1] The data used are the index numbers of area, production and productivity constructed by the Ministry of Food and Agriculture.[2]

Wide Fluctuations in Yields

During the seven years ending 1956-57, agricultural production showed a steady upward trend without significant dips. In the following eight years, however, the continued upward trend was marked by significant fluctuations from year to year. If we note the fact that output declined sharply during 1965-66 and 1966-67 and is expected to reach the highest peak during the current year (1967-68), it would appear that recent growth, far from reducing yield-instability, has been associated with increasing fluctuations.

Fluctuations in output may be decomposed broadly into fluctuations in sown area and in productivity per acre. Growth of area was significant during the period ending 1956-57, but slowed down considerably thereafter and almost stagnated after 1961-62. Productivity per acre, on the other hand, maintained its upward trend, the rate of growth being markedly higher in the later period; but the fluctuations in it were also quite marked, especially in the later period. The relative stability of growth in the first period is quite understandable because the contribution of cultivated area to growth was far more important. In the later period, however, growth became increasingly dependent on productivity, which fluctuates more than area. Moreover, productivity itself exhibited greater fluctuations in the later period. It is possible that most of the new lands brought under cultivation were marginal in the sense of being more vulnerable to the vagaries of monsoons.

It is also possible that the pattern of input-use was such as to induce yield-instability along with growth. In Punjab, Madras and Kerala, growth has been relatively stable; the coefficients of variation in productivity around the trend in these States are the lowest (Table 7.3). These States, nevertheless, differ significantly in respect of output growth: in Punjab and Madras the rate of growth of agricultural production between 1952-53 and 1964-65 was well above the national average (4.56 and 4.17 per cent, respectively, against the national average of about 3 per cent) while in Kerala it was 2.27 per cent.[3] The explanation for the relative yield-stability in these States seems to be that the proportion of irrigated to total sown area in Punjab and Madras is the highest among all States. This proportion is not so high, in Kerala but it accounts for the highest normal rainfall among all States, and the coefficient of variation in actual rainfall is also among the lowest (Table 7.3).

At the other extreme are the States like Gujarat, Madhya Pradesh and Maharashtra. The proportion of sown area irrigated is very low in these States. Even this meagre irrigation is provided mainly from uncertain sources such as tanks and wells (Table 7.3). The coefficients of variation in productivity around the trend for these States are among the highest in the country. These States also differ in regard to the rate of growth of production: Gujarat ranks second (4.55 per cent) in the country whereas Maharashtra and Madhya Pradesh are below the national average.[4]

In order to test and quantify more precisely the effect of irrigation and the pattern of irrigation on the amplitude of fluctuations in productivity, an attempt is made below to explain the State-wise differences in the coefficient of variation in

Table 7.3

Inter-State Variations in Agricultural Growth Rates, Fluctuations in Productivity and Proportion of Area Irrigated

(All Crops)

State	*Growth Rate of Productivity Per Acre (1956-57 to 1961-62)*	*Coefficient of Variation in Productivity (1956-57 to 1961-62)*	*Percentage of Sown Area Irrigated (1961-62)*	*Percentage of Irrigated Area Covered by Tanks and Wells (1961-62)*	*Normal Rainfall (Annual in Inches)*	*Coefficient of Variation in Actual Rainfall (1952-62)*
Gujarat	5.27	12.88	7.3	85.0	37.2	10.1
Bihar	6.15	8.80	17.5	28.3	50.4	2.7
Orissa	4.01	7.82	18.9	81.3	57.6	3.2
Madhya Pradesh	0.85	7.49	4.8	48.0	48.0	4.0
Maharashtra	2.47	7.41	5.8	73.8	34.8	4.0
Assam	2.77	3.94	25.5	NA*	99.6	1.3
Mysore	1.25	3.71	7.3	55.4	67.2	4.1
West Bengal	3.72	3.53	20.5	28.5	88.8	1.7
Uttar Pradesh	2.87	3.50	23.6	55.8	38.4	6.2
Andhra Pradesh	0.78	3.36	27.2	54.4	33.6	6.8
Rajasthan	2.18	2.37	12.3	65.4	19.2	8.8
Madras	2.66	1.60	42.2	61.2	38.4	1.7
Punjab	3.83	1.47	38.4	30.7	24.0	3.2
Kerala	0.14	0.94	20.3	13.0	116.4	1.8
India	**2.60**	**3.30**	**16.9**	**47.8**	—	—

* Not Available

Sources: (1) Government of India, Ministry of Food and Agriculture, "Growth Rates in Agriculture", 1964.

(2) Government of India "Statistical Abstract", 1962.

(3) Government of India, Ministry of Food and Agriculture, "Agricultural Situation in India", November 1964.

productivity between 1956-57 and 1961-62, in terms of the above two variables as well as a few other relevant factors. The regression coefficients have been estimated by ordinary least squares. The irrigation data relate to 1961-62. Inter-State differences in irrigation have not altered significantly during the period under study. Wells as a source of irrigation are a heterogenous category, representing assured as well as uncertain sources. Since data are not available according to the nature of these sources, all wells together with tanks and ponds have been treated as one category. It should be noted, however, that a reclassification would provide a better test of the hypothesis suggested here.

The proportion of sown area irrigated as a variable explaining inter-State differences in fluctuations is found to be significant. The results given in Table 7.4 show that the coefficient is fairly stable between various equations in respect of its sign as well as magnitude. The negative sign of this coefficient indicates that irrigation would have the effect of reducing the amplitude of fluctuations in productivity. Quantitatively, it would mean that if the proportion of sown area irrigated is increased by 1 percentage point (*e.g.* from 20 per cent to 21 per cent), then the coefficient of variation in productivity would be reduced by about 0.16. But irrigation by uncertain sources, *e.g.*, tanks and wells has an effect of increasing the fluctuations. The inclusion of this variable improved the total explanation (R^2) significantly. From this it appears very likely that a proper classification of minor irrigation sources according to whether they are assured or uncertain would improve the total explanation as well as the coefficients.

Another factor tested is the inter-State variation in the climatic conditions as reflected in the incidence of normal rainfall. But this variable does not turn out to be significant in explaining inter-State differences in fluctuations. This result

Table 7.4

Inter-State Variations in Fluctuations of Agricultural Productivity

Equation Number	*No. of Observations@*	*Ordinary Least Squares Regression Coefficients of the Explanatory Variables*				
		X_1	X_2	X_3	X_4	R^2
(1)	*(2)*	*(3)*	*(4)*	*(5)*	*(6)*	*(7)*
1	13	–0.18* (0.07)	—	—	—	0.37
2	13	–0.15* (0.06)	0.06# (0.03)	—	—	0.53
3	13	–0.16* (0.07)	0.07 (0.04)	—	–0.37 (0.81)	0.54
4	13	–0.17# (0.08)	0.06 (0.04)	–0.22 (0.51)	—	0.55

@ The State of Assam could not be included owing to the non-availability of data regarding area irrigated by tanks and wells.

* Significant at 5 per cent level.

Significant at 10 per cent level.

X_1= Proportion of sown area irrigated (1961-62).

X_2= Proportion of area irrigated by tanks (ponds) and wells to total irrigated area (1961-62).

X_3= Normal Rainfall (inches).

X_4= Size of the State (Geographical area in 1000 sq. miles).

should not be surprising because what are sought to be explained are not the inter-State differences in yield levels as such at any particular point of time or the variations in yields within each State over time, but the inter-State differences in the coefficient of variation in yields. The inclusion of the first two variables, *viz*., irrigation and pattern of irrigation holds constant the proportion of area dependent on rainfall. This may have subsumed, to certain extent, the impact of rainfall variability on the differences in productivity variations.

Another variable tried is the size of the State measured by its geographical area. The presumption is that a larger State is likely to be more heterogenous with respect to its climatic composition than a smaller State and, other things remaining equal, the fluctuations in productivity should be relatively smaller in a larger State because of the compensatory movements in productivity as between the divergent climatic zones within the State.[5] Although the sign of the coefficient for this variable is negative, it is not statistically significant and does not improve the total explanation substantially.

Of all the variables tested, therefore, two factors, *viz.*, irrigation and the pattern of irrigation explain significantly the inter-State differences in the levels of fluctuations.

That the recent growth of productivity per acre should have led to an increase in the amplitude of fluctuations would appear to be plausible in view of the negligible increase in the proportion of sown area irrigated and the steep rise in the inputs of labour, fertilisers and improved seeds. The proportion of sown area irrigated increased from 16.9 per cent in 1952-53 to 18.2 per cent in 1961-62.[6] That is to say, during the period of nine years it increased by only about 1.3 per cent or at the rate of 0.14 per cent of sown area per annum. Even according to the most optimistic estimates concerning the productivity of irrigation water, this would give a rate of growth in productivity of about 0.25 per cent per annum.[7] Since productivity per acre has grown in this period at 1.7 per cent per annum, it would appear that the bulk of the growth in productivity has been achieved through the use of inputs other than irrigation such as labour, fertilisers and improved seeds which, as discussed earlier, may have a built-in tendency to increase the amplitude of fluctuations. The number of agricultural workers has been growing at the rate of 3.6 per cent per annum during 1951-61,[8] as against an increase of sown area at the rate of 1.8 per cent during the corresponding period. The rate of increase in area declined sharply to 0.8 per cent per annum between 1956-57 and 1961-62 (Table 7.1).

Although the growth of agricultural work force may have been overestimated somewhat because of the differences in concepts and measurement between the two periods, these facts, nevertheless, suggest that there has been a steep rise in the labour inputs per acre. The consumption of nitrogenous fertilisers increased from about 60,000 tonnes N in 1950-51 to about 200,000 tonnes in 1960-61.[9] Again it was during this period that the improved seeds for foodgrains were introduced on a large scale, to cover about 50 million acres by 1960-61.[10]

Areas served by assured sources of irrigation constitute about half the total irrigated area (Table 7.3). Since only about 20 per cent of total sown area is irrigated, it would mean that assured sources account for only about 10 per cent of sown area. It is difficult to believe, therefore, that the increased inputs were concentrated in such limited areas. In view of the fact that even a State like Gujarat where the proportion of area irrigated is very small has shown a very high rate of growth in productivity indicates that the application of inputs was spread into uncertain areas also. Moreover, the application of fertilisers and improved seeds requires assured water for the maturity of the crop. In the year of good rainfall, therefore, these inputs can be applied with certainty even in areas served by minor irrigation sources. However, in the event of failure of rains the use of these inputs would be restricted in such areas. Therefore, one would expect the variation in yields to increase from year to year.

The agricultural work force has been increasing throughout the country, whatever be the differences in the extent and pattern of irrigation among different regions. It is known that the marginal product of labour is positive and significant for certain classes of cultivators, especially on the large and medium holdings.[11] The increasing subdivision of such holdings as a result of population pressure is an index of increasing intensity with which land is being cultivated. The increasing application of labour may, therefore, increase the amplitude of fluctuations in output because of the complementarity between the monsoons and the inputs.

The emphasis in the country has been shifting recently from the major and medium irrigation projects to minor irrigation sources. The Third Plan targets in regard to minor irrigation were fulfiled whereas the achievement in regard to major and medium irrigation fell considerably short of the targets.[12] Minor irrigation projects commend themselves for their potential in mobilising local resources and initiative and for their small gestation period. In a situation where food needs are pressing and where the pressure on public resources is increasing, it is understandable that there should be increasing emphasis on quick-maturing projects whose cost to the Government is much lower. But the above analysis suggests that the calculations regarding relative costs and benefits from different projects should also take into account the costs in terms of increased fluctuations in output associated with uncertain sources of irrigation. These may include the costs in terms of yield and price uncertainties, costs of procurement and storage of foodgrains and of their distribution between different regions and as well as over time.

With increasing emphasis on quick-maturing projects and modern inputs the fluctuations in output may increase further in many regions of the country, especially where the proportion of area irrigated is very limited. Since many of these regions are deficit in foodgrains even in normal years, their dependence on the surplus States is likely to increase further, especially in drought years. However, the position is not equally clear as to the future course of fluctuations in the aggregate foodgrains output for the country as a whole. The varietal improvements with a high capacity for fertiliser intake have opened up the possibilities of phenomenal increases in foodgrains output with limited land resources. If the resulting high profitability of food crops induces surplus States, which have large areas under assured irrigation, to divert more and more of their irrigated lands from non-food crops to foodgrains, then it may have a stabilising effect on the aggregate availability of foodgrains over time. But this would only shift the phenomenon of increasing fluctuations from foodgrains to non-foodgrains. Such a diversion of area is already evident in States like Punjab and Uttar Pradesh.[13]

Since fluctuations in output have been markedly lower in some of the surplus States, the expected growth of foodgrains output in these areas may have important implications for procurement, storage and distribution policies. Many of the so-called deficit States are deficit in foodgrains even in a normal year. Therefore, the problem is not so much of carrying stocks from good to bad years within each region — whether surplus or deficit — as of carrying stocks from the surplus regions to the deficit regions in good as well as bad years. This may also involve the building up of buffer stocks through procurement in surplus regions for meeting the needs of deficit regions in drought years. The increasing concentration of surpluses in the limited pockets of surplus States may call for a vigorous farm-price-support policy as the concentration of surpluses can depress the harvest prices even when the consumer prices in other States rule very high. The storage programmes — whether on public or private account — would have to become procurement-oriented and undertaken at a much larger scale than envisaged so far, because the storage requirements would be much larger with a spatial concentration of surpluses than when the surpluses are dispersed and find their way into pipelines.

If this prognosis is correct, then the need to support producers' prices may turn the surplus States into powerful advocates of free inter-State movement of foodgrains and a national food policy.

Notes

1. A similar method was used by Shephered for estimating the fluctuations in corn production in the United States. See Shephered, Geoffrey S: "Agricultural Price and Income Policy", p. 102.
2. Ministry of Food and Agriculture, Government of India, *Growth Rates in Agriculture*, 1964 and 1966.
3. Ministry of Food and Agriculture, Government of India, *Growth Rates in Agriculture, 1949-50 to 1964-65*, 1966.
4. *Ibid.*
5. This hypothesis is suggested by Schultz's findings in regard to corn yields in the United States. See Schultz, T.W., "The Economic Organisation of Agriculture", pp. 203-04.
6. Ministry of Food and Agriculture, Government of India, (i) *Economic Survey of Indian Agriculture 1960-61* p. 24; (ii) *Agricultural Situation in India*, November 1964, pp. 771-72.
7. Hanumantha Rao, C.H. "Agricultural Growth and Stagnation in India", *The Economic Weekly*, February 27, 1965.
8. Government of India: (i) "Census of India", Paper No. I of 1962, 1961 Census Final Population Totals, pp. 402-3; (ii) "Census of India", 1961, Vol. I, India, Part II-A; (iii) "Union Primary Census Abstracts", p. LVII.
9. Ministry of Food and Agriculture, Government of India, *Agricultural Development: Problems and Perspective*, 1965.
10. *Ibid.*
11. Hanumantha Rao, C.H. *Agricultural Production Functions, Costs and Returns in India*, Asia Publishing House, 1965.
12. Government of India, Ministry of Food and Agriculture, *Report of the Foodgrains Policy Committee*, 1966.
13. In Punjab the indexes of area under rice and wheat registered a steady increase reaching 172 and 121, respectively, in 1964-65 (1956-57=100) whereas the indexes for cotton and sugarcane fluctuated considerably during the same period. Similarly, the area index of rice recorded a steady increase in Uttar Pradesh as against considerable fluctuations in sugarcane area. See, Government of India, Ministry of Food and Agriculture, "Growth Rates in Agriculture", 1966.

Economic and Political Weekly, Annual Number, January 1968.

8

Fluctuations in Foodgrains Output, 1950-51 - 1979-80: Implications for Policy

The severity of drought this year (1979-80) leading to a significant fall in foodgrains output has brought to the fore the question whether the 30 years of development planning has contributed to reducing the vulnerability of foodgrains output to the vagaries of monsoons.

The year-to-year variations in agricultural output are caused by a combination of two factors: (a) the changes in inputs and technology, *e.g.* irrigation, fertilisers, use of high yielding seeds, *etc.*; and (b) the changes in weather *e.g.* rainfall, temperature, humidity, *etc.* If there is an increase in the application of inputs, particularly the technologically new inputs, under rainfed conditions including irrigation from rainfed sources like tanks, there may be a spurt in yield in a year of good rainfall, because of complementarity between such inputs and moisture. In a year of deficient rainfall, however, the yield may decline steeply. Thus, the amplitude of fluctuation in output may increase with the increase in inputs under such conditions. In areas served by assured sources of irrigation, on the other hand, the growth of such inputs may lead to a reduction in fluctuations. Therefore, the objective of growth with stability is best achieved when the growth of inputs and the improvements in technology are accompanied by the development of assured sources of irrigation.

In a period characterised by a sustained increase in inputs, one should, in general, expect the positive year-to-year deviations in output to be more marked than the negative deviations. This is because in a year of good weather (rainfall), the effect of inputs as well as weather would be positive. In a year of bad weather (deficient rainfall), however, the normal increase in inputs can be expected to provide a cushion and reduce the adverse effect of weather. The net impact would, however, depend upon the relative strength of each of these factors.

Year-to-year variations in the total foodgrains output for the country as a whole for 30 years from 1950-51 to 1979-80 (Output of 119 million tons is assumed for the current year) have been plotted in Figure 8.1. These deviations represent the percentage change in the output in each year when compared to the preceding year.

There are in all 17 positive deviations and 13 negative deviations. The average of all these positive (percentage) deviations is +9.98 which is significantly higher than the average of negative deviations which comes to -5.86. For reasons mentioned above, in a period of growth, the positive deviations can be expected to be stronger than the negative deviations. However, it appears from Figure 8.1 that there is no clear tendency for these deviations, whether positive or negative, to decrease or increase over the last 30 years. The available studies show that except for the brief period in the 1960s, when the growth of foodgrain output had slowed down and in the late 1960s and early 1970s when there was a marked improvement in the foodgrains output consequent on the breakthrough in wheat output, there is no evidence of either acceleration or deceleration in the rate of growth of foodgrains output in the country as a whole over a long period. This is consistent with the lack of any trend in the positive deviations in Figure 8.1. The absence of any declining trend in the negative deviations indicates that agricultural development has failed to reduce the adverse impact of weather, so that, in proportionate terms, the foodgrains output is about as susceptible to weather now as in the beginning of economic planning in India.

Figure 8.1

Foodgrains (All-India)
Year-to-Year per cent Deviations

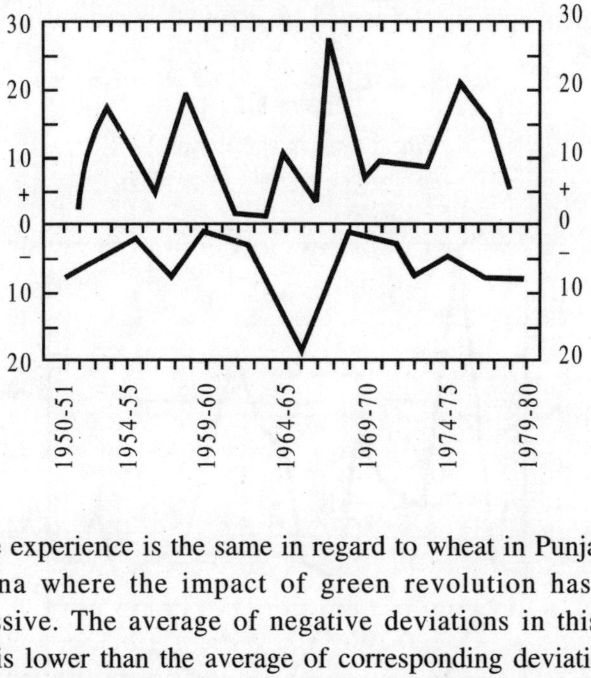

The experience is the same in regard to wheat in Punjab and Haryana where the impact of green revolution has been impressive. The average of negative deviations in this case (-6.7) is lower than the average of corresponding deviations in the case of wheat for the country as a whole (-9.4), and the average of positive deviations (+16.1) is higher than the average of corresponding deviations for the country (+13.1). This is understandable in view of the fast rate of growth of wheat output in Punjab under conditions of assured sources of irrigation. However, as is clear from Figure 8.2, there is no tendency for the negative deviations to become smaller over time. Although the bulk of the cultivated area in Punjab and Haryana is served by assured sources of irrigation like canals and tubewells, as much as 90 per cent of the area in this region accounts for low rainfall upto 750 mm. The variability of rainfall is generally greater in the low rainfall regions and, according to a study by S.K. Ray, the output response to a unit change in rainfall is generally higher at low rainfall levels than at high rainfall levels. In view of these two factors, the variability of output can be expected to be greater in the low rainfall regions. This factor cannot be ignored altogether even for regions adequately served by assured sources of irrigation.

Figure 8.2

Wheat (Punjab and Haryana)
Year-to-Year per cent Deviations

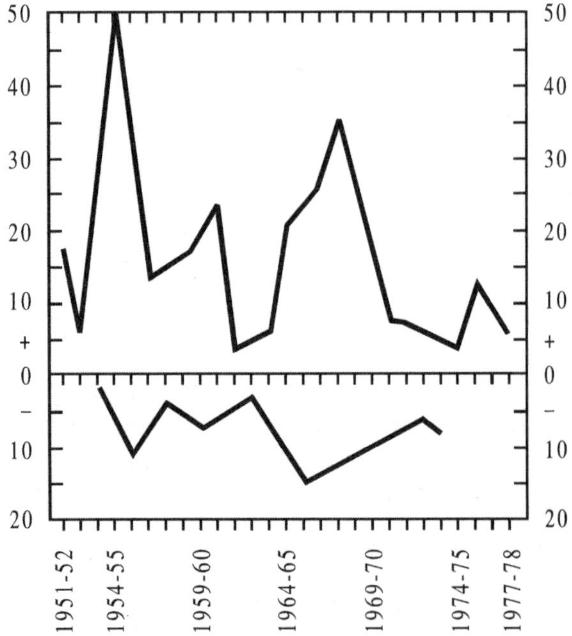

The case of rice in Tamil Nadu, presented in Figure 8.3, seems to be quite different. The average of negative deviations (-12.69) is higher than that of positive deviations (+11.83).

There is also a tendency for the negative deviations to increase over time. About 83 per cent of area in Tamil Nadu falls under medium rainfall zone (750 to 1150 mm). The proportion of area irrigated in Tamil Nadu is much lower than in Punjab and Haryana. Further, bulk of the irrigated area in Tamil Nadu is served by uncertain sources like tanks and wells which are dependent on rainfall.

Figure 8.3

Rice (Tamil Nadu)
Year-to-Year per cent Deviations

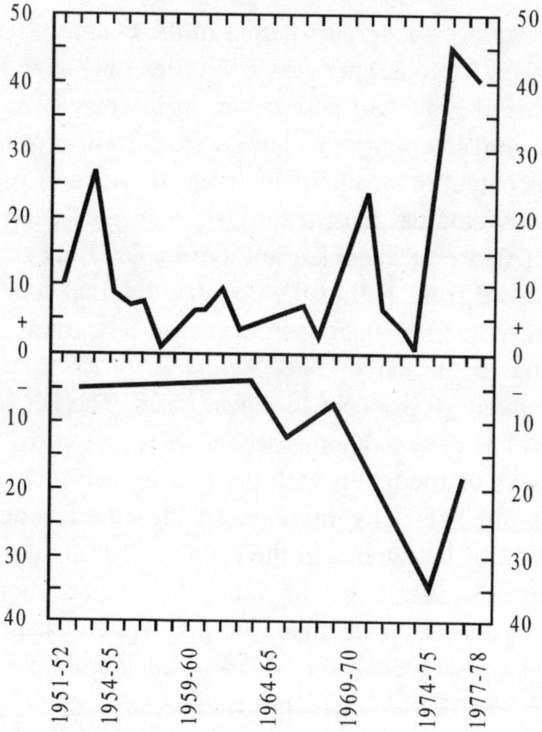

The above evidence suggests that there is no unique relationship between the growth of output and the degree of instability. Degree of instability depends on the pattern of growth, particularly on the magnitude and pattern of irrigation, given the weather conditions. About one-fourth of the cultivated area in the country is now under irrigation. Out of the remaining irrigation potential of about one-fourth of cultivated area, about two-thirds is accounted by the relatively assured sources of irrigation such as major and medium irrigation projects to be undertaken through public investment. This together with the exploitation of the relatively assured groundwater resources in the vast Indo-Gangetic Plains, holds the prospect of a significant reduction in the vulnerability of foodgrains production to the vagaries of monsoon.

The growth of foodgrains output in the country – estimated to be between 2.6 to 3.0 per cent – has been only slightly higher than the rate of growth of population. As discussed above, there has not been any marked change in the susceptibility of foodgrains output to weather. However, the stock of foodgrains with the Government has steadily risen in the last few years reaching a figure of about 20 million tonnes. Bulk of this has been procured from within the country. Internal procurement which constituted less than one per cent of foodgrains output in the country in the early 1960s has steadily risen reaching a figure of about 10 per cent in recent years. The net import of foodgrains has recorded some decline in recent years. The rise in the stocks of foodgrain with the Government indicates that there has not been any increase in the average per capita consumption of foodgrains in the country. Indeed, according to a study by J.S. Sarma and Shyamal Roy of the International Food Policy Research Institute, the per capita consumption of cereals has shown a decline from 144.0 kilograms/year in 1960-62 to 142.5 in 1975-77. The per capita consumption of pulses declined from 24.0 kilograms to 16.4 kilograms during the corresponding period.

The rise in the stocks with the Government from internal procurement and the decline in the per capita consumption of foodgrains is traceable to the pattern of agricultural development which has been characterised by regional disparities as well as disparities between large and small farmers. The rate of growth of foodgrains output has been very high in certain developed regions like Punjab, Haryana, West U.P. and certain prosperous pockets of Tamil Nadu and Andhra Pradesh where the income elasticity of demand for foodgrains is very low and where, therefore, the bulk of the increase in output has been marketed, especially to the public procurement agencies. For example, Punjab, Haryana and U.P. together accounted for 60 per cent of the total foodgrains procurement by the Government in 1975-76 as against only about 27 per cent in 1964-65. These states, which accounted for only about 13 per cent of the rice output in the country in 1975-76, contributed as much as 40 per cent to the rice procurement by the Government. Similarly these three states accounted for 37.4 per cent of total output of wheat in the country in 1975-76 and contributed as much as 88.5 per cent of wheat procured by the Government.

It is possible, therefore, that in large parts of the country where the rate of growth of foodgrains output has been much lower than the all-India average, the per capita consumption of foodgrains in the rural areas, particularly in the case of the rural poor, has declined because of a decline in per capita income. The decline in the average per capita consumption of foodgrains in the country despite an increase in per capita income and the stability of relative prices of foodgrains over this period can be explained by the marked shift in the distribution of rural incomes in favour of the developed regions where the income elasticity of demand for foodgrains is very low.

The increase in the stocks of foodgrains with the Government has enabled the Government to respond quickly to the shortages in the urban areas in bad years. Thus, unlike in the first two decades of planning, we are now in a position to meet the requirements of foodgrains in urban areas from internal sources. But in the rural areas, the problem is essentially one of low purchasing power for the poor in regions where the growth of agriculture has been slow. This position is aggravated in the drought years as in the current year. All these factors seem to have contributed to a decline in the urban-rural differential in the prices of foodgrains.

The problem is, therefore, essentially one of raising the purchasing power of the rural poor. Two alternatives are open to achieve this objective. The first is to broaden the agricultural growth involving increasingly the lagging regions and small farmers. This would require early exploitation of the remaining irrigation potential from major and medium irrigation projects through public investments. The regional distribution of these water resources is such that their exploitation would result in the reduction of regional disparities in agricultural development. For example, the inter-state variation in this potential is lower than the variation for the potential exploited so far from such sources. Similarly, the inter-state variation in regard to exploitable groundwater potential is lower than the variation in regard to the groundwater exploited so far. Land reform measures insofar as they contribute to enlarging the area held by small and marginal farmers will raise their incomes and purchasing power. Credit reform and the supply of inputs benefiting these sections would also contribute to this objective.

Such a strategy would generate more employment for the rural poor than the strategy of unbalanced regional development. This is because although the labour from the poorer regions would migrate to the developed and high wage pockets, for employment, there are economic and psychological costs of migration and, in view of the high wage rates in the developed regions, the total employment generated would be smaller than if the growth rate is stepped up in the labour-abundant and low-wage areas.

All these measures would contribute to reducing the fluctuations in output. Agricultural output resulting from assured sources of irrigation as well as increased labour use constitutes a stable component of output growth. Increase in cropped area resulting from cropping intensity which is usually higher among the irrigated and the small farms also contributes to stability in output growth. Spatial differences in weather are quite significant so that the fall in output in any year in certain regions is neutralised to some extent by the rise in output in other regions. Therefore, regionally diffused agricultural growth is likely to lend greater stability to the growth in foodgrains output for the country as a whole than if growth is concentrated in a few regions.

The second alternative is the one which the country has actually been experiencing. This strategy would enable the government to accumulate stocks of foodgrains in sufficient quantities even if the overall growth of output is lower. The task would then be to distribute these foodgrains among the rural poor by raising their purchasing power through Rural Works Programme and through Food-for-Work Programme in drought years. The Employment Guarantee Scheme of Maharashtra is an example of this kind. The Government has undertaken a massive Food-for-Work Programme this year involving the distribution of 3.5 million tonnes of foodgrains. Such programmes entail massive administrative effort and encounter the most difficult problems of management. The distance between the decision makers at the highest level and the beneficiaries at the lowest level is, perhaps, the greatest in this case involving a large number of intermediaries and substantial leakages associated with any such programme. The capital assets actually created and the purchasing power actually generated for the poor is likely to fall very much short of the expectations. In view of these inherent difficulties, the first strategy to generate employment and income for the rural poor in the process of agricultural growth itself is likely to be more efficient. However, the second alternative of Rural Works Programme and Food-for-Work Programme would also be necessary to the extent that we fail in the first strategy.

The Economic Times, February 8, 1980.

9

Unstable Agriculture: Lessons from Current Drought (1987-88)

Agricultural output has grown at the rate of about 2.7 per cent per annum. But this growth has been characterised by marked fluctuations in output from year to year around a rising trend. What is more, there is no evidence, as yet, of a decline in the amplitude of these fluctuations over a period of time. Indeed, a number of studies point to a rise in this variability in the post-green revolution period.

However, there has been a belief in certain policy-making circles that agricultural production has been showing resilience as evidenced by smaller dips in unfavourable years. But this optimism has been based on the experience in the eighties before the onset of the current drought (1987-88). In the very nature of things, a judgement on the variability of agricultural output has to be based on the experience over a reasonably long period. Weather-wise, no two bad years are comparable in respect of the severity of drought. The drought of exceptional intensity, such as in the current year or 1979-80 or 1966-67, is of rare occurrence which can be captured only when a reasonably long period is considered for study. The same is true of exceptionally good years like 1983-84, which can give rise to undue optimism if a short period is considered.

Whereas the sensitivity of agricultural output to weather seems to have increased over a period of time, our ability to cope with the consequences of drought has undoubtedly improved. Droughts of much less severity than the one this year or even of localised nature used to result in widespread famine and misery in the pre-independence period. Large buffer stocks of foodgrains with the government, an elastic transport system, improved distribution network, and above all, an easy and free flow of information, together with the responsiveness of the government in a democratic polity to the distress situations, have contributed to relieving misery from droughts, to a considerable extent.

However, the costs of output instability to the economy are considerable. Measures to reduce instability in production up to a point could well prove to be less costly to the cconomy and polity than the costs of maintaining huge buffer stocks of foodgrains and undertaking distribution of grain as well as operations for relieving the shortage of drinking water and fodder for animals. Since agriculture is a highly labour-intensive activity, a drought of severe intensity substantially erodes the incomes of wage labour and marginal farmers, especially of those in the dry or rain-red pockets who constitute the poorest of the poor. The costs of providing employment opportunities to these poorer sections through public works to compensate for the decline in their purchasing power is enormous. Imparting stability to agricultural production should, therefore, receive the highest priority in agricultural planning.

Such a policy has to be based on an adequate understanding of the factors responsible for the observed rise in the amplitude of fluctuations in agricultural output. Of the two components of agricultural growth, *viz.*, cultivated area and yield per acre, the latter is far more sensitive to weather than the former. In the early phase of planning or the pre-green revolution period, expansion in cultivated area as a source of growth was much more important than in the post-green revolution period, starting from the late sixties. The declining importance of area expansion, which is a relatively stable component of growth, and the rising significance of yield — an unstable component — explains to a certain extent the increase in the amplitude of fluctuations in agricultural output. Further, with area ceasing to expand significantly, there is a greater competition for the given area between foodgrain and non-foodgrains. Thus, when rainfall is favourable more area is allocated to foodgrains, whose yields per acre also rise. The converse happens in the event of failure of rainfall. This is particularly true of the period following the adoption of new technology which is crucially dependent on moisture availability. These sympathetic movements between area and yield for foodgrains contribute to reinforcing the fluctuations in their output.

What is more important is that the variability in yield per acre itself has increased in the post-green revolution period. This is attributable basically to the complementarity between new technology like high-yielding seeds and fertilisers on the one hand, and water on the other. In a year when soil moisture is adequate and the water table favourable on account of good rainfall, yields of foodgrain are pushed up significantly, but in a year of deficient rainfall, these yields are pushed down, thus widening the year-to-year differences in yields.

Although about 30 per cent of cultivated area is under irrigation now, as against only about 17 per cent in the early years of planning, a good part of this irrigated area is dependent on tanks and wells which are themselves dependent on rainfall to a considerable extent. Whereas the tubewell revolution has made irrigation 'controlled', in the sense that unlike canal irrigation, farmers can regulate the timing and number of irrigations for crops, this irrigation has nevertheless remained 'uncontrolled', in the sense that it is sensitive to rainfall, especially in areas where the water table is not favourable. Since high yielding varieties of seeds were cultivated and fertilisers applied intensively under the so-called 'controlled' irrigated conditions, the annual variability in yields was bound to increase.

With the advent of the green revolution, there has been a greater concentration of output in certain regions in the northern and western parts of the country. Wheat and rice in Punjab, Haryana and western Uttar Pradesh, *bajra* in Gujarat, *jowar* in Maharashtra, and *ragi* in Karnataka have shown impressive results. In fact, the rate of growth of agricultural output in quite a few of these states has been above the national average. These states are characterised by low rainfall which is also highly variable. Although irrigation is extensive in Punjab, Haryana and western Uttar Pradesh, rainfall supplements irrigation, not only directly but also indirectly by influencing the water table. This is particularly true for *Kharif* crops including rice. In Gujarat, Maharashtra and Karnataka, despite a low level of irrigation and low and variable rainfall, fertiliser application is widespread because of better institutional framework, stronger resource position of farmers and higher overall development. With growth, therefore, variability of output has also increased in these areas. In a large country like India with its diverse agro-climatic conditions, regional dispersal of output growth is likely to ensure a more even flow of output for the country as a whole because of compensatory movements in yields between different regions. Concentration of output growth in certain limited pockets, on the other hand, tends to accentuate fluctuations in output.

The factors discussed above have, on the whole, contributed to increasing instability in the output of *Kharif* cereals including rice. In the case of *rabi* cereals, consisting essentially of wheat, on the other hand, instability in output has declined significantly. This is because the direct influence of monsoons on *rabi* crops is limited. *Rabi* crops are generally grown under assured irrigated conditions. Unlike rice which is grown largely in *Kharif*, the water requirements for wheat are comparatively less which explains its lower vulnerability to rainfall variations.

The above analysis suggests, first, that the policies for balanced development of agriculture between different regions would contribute to reducing instability in the flow of output at the aggregative or national level. Second, the development of irrigation from assured sources such as major and medium irrigation projects holds the key to reducing instability in agricultural output. It has been estimated that when the available potential from these sources is fully exploited, the regional disparities in irrigation will narrow down considerably. The gap between plan outlays and actual expenditure has been greater in respect of irrigation during the Sixth Plan period. Irrigation is essentially the responsibility of the states which account for about 90 per cent of plan outlay for this sector. Both paucity of resources, particularly in the case of less developed states, and the low priority assigned to this crucial sector are responsible for slow growth of irrigation.

Third, exploitation of groundwater in high rainfall regions such as in the eastern Gangetic belt where the water table is favourable is likely to impart stability both by reducing variability in *Kharif* output and by expanding area under *Rabi* crops through multiple cropping. At the same time, policies for water release and charges for water use in water-scarce regions should be such as to encourage its extensive use by spreading it thinly with a view to maximising output per unit of water. This would contribute to equity as well as stability.

Our plan achievements in the recent period in respect of increases in gross cropped area through multiple cropping have fallen short of expectations. This is due in part to the slow exploitation of groundwater potential in regions like the eastern Gangetic belt, where cropping intensity is low despite the predominance of small farmers and abundant availability of labour. This is also explained in part by the generally capital-intensive nature of new technology which is particularly suited to the large farmers who are unwilling to absorb more labour necessary for multiple cropping and who, instead, seek to maximise output on the area already under cultivation through the intensification of modern inputs.

Policies for increased labour absorption in agriculture are, therefore, likely to lend stability to output flow through multiple cropping and through improved water management. In this context, consolidation of holdings can provide the necessary incentives for exploitation of groundwater on private account. Agricultural research can play a major role by evolving cropping systems which enable the use of a piece of land for the maximum duration in a year and by evolving techniques for water-harvesting and moisture conservation which enable the farmers to withstand the drought.

The proportion of population below the poverty line is much higher than the national average in the dry or unirrigated areas where rainfall is low and is highly variable. The poor are found to migrate from such areas whenever the crops fail. Therefore, measures to reduce instability in agriculture are likely to have the greatest impact in terms of raising the levels of living of the poor.

Ecological degradation on account of denudation of forests has increased our vulnerability to droughts. Loss of soil cover and depletion of soil moisture on account of deforestation has led to fodder scarcities even in normal years. Scarcity of drinking water has been accentuated on account of the water table becoming low because of run-off and insufficient percolation of water. Fuel-wood crisis which is being felt even in normal years is accentuated further in drought years. Vulnerable groups in rural areas, particularly women and children, who procure fodder and fuel-wood and fetch drinking water, are most affected by these scarcities. Afforestation programmes, which have received some attention in the recent period, need to be stepped-up for raising our capacity to withstand droughts.

Programmes for improved water management and afforestation cannot succeed without the active involvement of the people and their representative institutions at the grassroots level. Harnessing the creative energies of our people through group endeavour constitutes one of the biggest challenges in our fight against the vagaries of nature. Voluntary agencies have been assigned a major role in the Seventh Plan for mobilising such efforts. Such arrangements need to be continuously reviewed with a view to overcoming the bureaucratic inhibitions in enlisting the cooperation of non-official agencies and peoples' institutions.

No less important is the need for a change in attitude on the part of the policy-making circles who have tended to take an extremely short-term view regarding the performance of agriculture. The fluctuating impact of weather on policy-making has been quite visible.

A succession of two of three good harvests in the past has invariably prompted the policymakers to take credit for "good performance" and to assert that stability in agricultural output has been achieved and to prepare plans for entering the export market.

This has lulled them into complacency in regard to the much needed efforts to increase irrigation facilities and to conserve moisture. On the other hand, the fall in output as a result of drought is treated as an exceptional and rare phenomenon and the efforts are, therefore, directed essentially towards, short-term measures for providing relief.

'National Water Grid' is a desirable goal but is a distant one owing to its large demand on resources including power. The speedy completion of the innumerable ongoing major and medium irrigation projects through adequate financial provisions and tight monitoring together with the development of forestry, provide a practical solution for mitigating the incidence of droughts in the foreseeable future.

The Economic Times, October 19, 1987; Also in *Economic Journalist*, September, 1987.

10

Prospects for Crop Insurance in India: Some Issues for Consideration

Area Approach

There seems to be a consensus that we in India should follow the area approach for determining the premium rates as is done in the U.S.A. and certain other countries. However, there seems to be some difference of opinion in regard to whether the same Area Approach is to be followed for determining the indemnities. In the U.S.A., indemnities are calculated for each individual farm on the basis of the realised yields on the farm. The moral hazards may turn out to be serious if such a method is followed under Indian conditions. Apart from the possibility of serious under-statement of realised yields, farmers may lose incentives for investment beyond a point. For instance, if farmers are indemnified to the extent of two-thirds of the difference between the actual farm yields and 75 per cent of the normal area yield and if the expected area yield in any year is 66.6 per cent of the normal yields, the farmers may restrict investment so as to achieve only about 50 per cent of the normal area yield, as they would any way get the remaining 16.6 per cent of the normal area yield by way of indemnity. Besides, it is far more difficult to get dependable yield data relating to individual farms in India than in the U.S.A. where many farmers maintain records covering fairly long period and there are several independent checks and balances for ensuring the accuracy of information. Moreover, only about 5 per cent of the total population is directly dependent on agriculture in the U.S.A. and the number of farms to be dealt with is restricted much further owing to the voluntary character of the programme and to a high degree of selectivity in respect of regions as well as persons practised by the insurance authorities there. In India, the number of farms to be dealt with would be unmanageable in any homogenous region especially if the programme is to be compulsory and hence the moral hazards and the administrative costs may far outweigh the possible benefits from the programme. Area Approach for determining premia *as well as* indemnities seems, therefore, inevitable in India if Crop Insurance is to be introduced at all.

Method for Determining Premium Rates

The essence of the various methods followed in many countries including those proposed in India for rate-making is that for any relatively homogenous region the value of indemnities payable over a fairly long period should approximately equal the amount of premium. The choice of the particular method would depend upon how realistic its assumptions are considering the prevailing yield data. Therefore, in the absence of long-period data on yields for relatively homogenous regions, it is difficult to establish the superiority of any one method over the others. The method suggested by the Indian Agricultural Research Statistics (IARS) assumes a normal distribution of yields and chooses the values of the parameters of this distribution, *viz.*, mean and variance from particular sets of data relating to six years. For the rates so worked out to be realistic one has to make sure that the behaviour of past data does not seriously violate the assumption of normal distribution and that the values of the parameters chosen are close to those revealed by the data. In the United States where a variant of the normal curve method is followed, only those counties are selected for the crop insurance scheme where the yield data are found to be normally distributed and the values of the parameters chosen are those revealed by such data. It would, therefore, be useful to examine the available data for homogenous regions in India to ascertain whether the assumption of normal distribution is realistic and if so over how long a period. Premium rates would no doubt be independent of the length of the period for which the assumption of normal distribution holds true but if the period is unduly long, say more than 20 years, even though the scheme would be self-financing over this period, government may have to step in with considerable financial resources for the payment of indemnities in certain years and thus lock up its resources for long periods.

Unlike the IARS method, the method followed by the Agricultural Prices Commission (APC) in their April 1970 Report makes no assumption about the shape of the distribution but shares the assumption common to all other methods *viz.*, the past pattern is likely to be substantially repeated in future. One merit of the APC method is that it operates with the actual data covering a reasonably long period and concentrates on the yield deviations below 75 per cent of the average yields. The IARS method, on the other hand, is subject to the limitation that the variance may sometimes be unduly influenced by a few large deviations above the mean. However, in such an event, the premium rates would only be overstated and not understated so that the resulting rate structure is likely to be in keeping with the financial viability of the scheme. Thus, the problem in regard to rate-determination concerns not so much the choice of the method on *a priori* grounds as the knowledge regarding the behaviour of yield data over reasonably long periods.

The Characteristics of Insurance

The contemplated crop insurance scheme is designed to collect from each farmer enough money every year so as to meet a specified part of his yield losses in bad years. Under an area approach (in respect of the fixation of premia as well as the payment of indemnities) all farmers pay similar premium rates and get similar amounts (per acre) as indemnities in bad years. Thus over a period for which the assumptions of the particular method followed hold true, each farmer receives an amount equal to what he has paid. This amounts to evening out the income of each individual over time. Even in the United States where indemnities are determined on the basis of individual yields, enough care is taken to see that each individual farm and the homogenous region receives in the long-run what it contributes and that no particular group benefits significantly at the expense of others. Thus regions and individuals showing losses frequently are weeded out of the programme and rebates on premiums are given to those who do not show losses continuously over a period and the rebates are raised significantly with the increase in the length of such period.

Such schemes, especially the one contemplated for India, cannot be called insurance in the strict and the generally accepted sense of the term. According to Alfred Manes "the essence of Insurance lies in the elimination of uncertain risk of loss for the individual through the combination of a large number of similarly exposed individuals who each contribute to a common fund premium payments sufficient to make good the loss caused any one individual" (*Encylopaedia of the Social Sciences,* Vol. 8, p.95). Although the individuals thus combined are similarly exposed to the risk and therefore contribute to the common fund equally, the actual incidence of loss may not be the same for all individuals so that the indemnities for those who actually suffer losses would exceed the premium amounts paid by them. Thus the economic function of insurance is 'the substitution of a known small loss (*i.e.*, premium payable) for an uncertain large loss..... the fact that the payments of all participants aid each participant in case of need is an essential characteristic of insurance' (*Ibid.*, p. 97). The losses suffered by a few individuals are thus spread horizontally over a large number of individuals in any meaningful insurance scheme. Insurance becomes attractive because the premium rates charged would be a small fraction of the extent of (uncertain) loss for any individual.

The actuarial characteristics of crops in any homogenous agricultural region may not be such as to satisfy the above principle. Crop losses when they occur would be fairly widespread and it is probable that individual yields in any bad year would be normally distributed. Such an assumption seems to be implied in the U.S.A. method of calculating premium rates. The actuarial considerations facing fire, life, and cattle insurance, on the other hand, are radically different. To take an instance, the death loss per 100 horses insured by an Association in the U.S.A. during 1936-41 was around 3.9 for all ages (P.K. Ray, *Agricultural Insurance*, 1967, p. 258). Similarly, over a period of 38 years during 1892-1930 the death rate per 100 horses reported by the Live Stock Insurance Society in Sweden was a little less than 3 for all ages. (*Ibid.*, p. 259).

Farmers are no doubt similarly exposed to the uncertainty of crop losses in any homogenous region, but the difficulty from the standpoint of insurance seems to be that when they suffer losses bulk of them do so nearly to the same extent (per acre), so that each one has to pay largely for oneself instead of others largely paying for oneself. This difficulty can be obviated to some extent if different agricultural regions are consolidated into one group and are required to pay similar premium rates irrespective of the differences in yield variability. But this procedure violates the basic principle of insurance that individuals combined should be similarly exposed to the risk. Where risks are markedly dissimilar, as they obviously are between different agricultural regions, uniform rates would have the effect of a tax on the regions showing lower yield variability and would therefore be resented as being iniquitous. Area approach is therefore inevitable and hence the contemplated scheme assumes the character of a kind of income stabilisation scheme for farmers rather than Crop Insurance.

Demand for Crop Insurance

Since the premium rates will have to be sufficiently high in order that each farmer pays for himself in the long-run, it is doubtful whether such a scheme of crop insurance would be acceptable to the large majority of farmers. As it is, farmers have at their command several means of stabilising their incomes. They do save a larger proportion of their income in good years and invest in capital assets such as farm cattle, machinery, irrigation, *etc.* and in consumer durables including textiles. They are also known to invest in precious metals such as gold and silver. They draw upon these assets in bad years either by selling some of them or by offering them as collateral for obtaining loans. Also, they can restrict expenditure on consumer durables in bad years.

The contemplated scheme provides insurance against the loss of current agricultural income and not against the loss of capital assets which are a major source of income as well as of its stability. Bad weather may be damaging to crops but may leave capital assets largely intact. Floods damage the crops but very often enhance the fertility of soils. However, the loss of capital assets when it occurs would amount to a multiple of current income. Demand for the insurance of capital assets can, therefore, be expected to be greater than for the insurance of crops particularly because the premium rates can be lower owing to the possibility of horizontal spread of risk in the case of capital assets. This may explain why compulsory insurance has worked out successfully in the Emergency Risks (Goods and Factories) Insurance. It is difficult to conclude from this experience, that the introduction of Crop Insurance on compulsory basis should not be very difficult.

Crop Insurance based on area approach may not prove to be effective even as a means of stabilising income. The yield data circulated by Mr. B.N. Kapre indicate that even in a good year like 1967-68, a significant proportion of plots in different states showed yields less than 20 per cent of average yields. Since farmers are indemnified on the basis of average loss for the area, such farmers who sustain severe losses may get only a nominal relief whereas those showing high yields may get equal benefit. Similarly, large farmers who can better withstand crop losses get support on par with small farmers. Moreover, the variability in the overall productivity is much smaller than the yield variability of individual crops, owing to the compensatory movements among the latter. Since the contemplated scheme is crop-based, the reliefs provided on the basis of individual crop losses may not always be necessary for income stabilisation except in years when all the major crops suffer losses.

In view of the above limitations it is doubtful whether the farmers would be enthusiastic about crop insurance as a means of income stabilisation if they grasp the scheme fully, especially when the nationalised banks and other credit institutions are providing new opportunities and incentives for the farmers to save in good years so that they can help themselves in bad years. Assets in the form of saving deposits even when they are made compulsory may be valued more by the farmers than the uncertain reliefs in the form of indemnities against crop insurance policies which can at best be regarded as highly illiquid assets.

Relative Efficiency of Prevention and Assumption of Risks

Insurance represents the assumption of risks through consolidation and is to be preferred only if the costs of assumption are lower than the costs of prevention of risks. Under the contemplated Crop Insurance Scheme the risks facing any farmer are, in effect, assumed entirely by the individual concerned. It would be necessary to examine whether such a method of stabilising one's income is more efficient when compared to the measures to reduce yield variability by investing the same funds. For a traditional agriculture which has taken a start for development, a good part of the yield risks are such as can be prevented or reduced through developmental investments. Investments in irrigation and pesticides reduce yield variability apart from increasing productivity. In India, irrigation bears a strong positive relationship to productivity and inverse relationship to yield variability. According to a study, about 55 per cent of inter-state difference in yield variability was explained by the irrigation factor alone (See Chapter 7).

A study by the National Council of Applied Economic Research shows that a rupee invested in pesticides yields in general a return of three rupees (NCAER, *Pesticides in Indian Agriculture*, 1967). The following simple numerical model taking the example of investments in pesticides illustrates how a rupee allocated to the prevention of risks is likely to be more efficient than its allocation to Crop Insurance. It shows how apart from reducing income variability more effectively than crop insurance, the investments in the prevention of risks increase farmer's income.

These figures illustrate that even when yield falls sharply from Rs. 100 to Rs. 50 due to the incidence of pests, crop insurance scheme can provide relief to the farmer only to the extent of Rs. 4.4 whereas if the farmer invests the same amount in pesticides the relief he can provide rises fourfold (*i.e.* Rs. 17.6) and his average annual income would be higher by Rs. 4.4. Unlike insurance, the prevention of risks improves the climate for investment in agriculture by reducing the yield variability and raising the expected yield. Insurance becomes economical

Year	Yield (Y) Rs.	Indemnity (Rs.) =2/3(.75 Y-Y)	Premium (Rs.)	Income (Rs.) (1+2-3)	Investment in pesticides (Rs.)	Return 1:3 (Rs.)	Income (1-3+6) (Rs.)
	(1)	(2)	(3)	(4)	(5)	(6)	(7)
1	100	—	2.2	97.8	—	—	97.8
2	50	6.6	2.2	54.4	6.6	19.8	67.6
3	90	—	2.2	87.8	—	—	87.8
Average (Y)	80			80.0			84.4

only when the possibilities of preventing risks are minimal so that the cost of prevention is higher than that of assumption.

Experience of U.S.A. and Japan

The experience of United States and Japan in respect of crop insurance is sometimes cited in favour of its introduction in India. However, the experience in these countries has not been encouraging and even if the prevailing schemes in these countries are considered justifiable, their relevance for India is not straightforward as the two situations are highly dissimilar.

Crop Insurance in the United States has now been reduced to an experimental programme and is very much restricted in coverage as compared to its earlier phase. However, the programme has proved to be self-financing in the last decade. But the premium rates still constitute, on an average, 6 per cent of the value of crops insured. A similar percentage for India would mean a three to four times the amount of land revenue collected at present and would therefore be beyond the capacity of the farmers in regions of low productivity and high yield variability. Even in a rich country like Japan about 50 per cent of the value of indemnities is subsidised by the government.

It is probable that a good part of the prevailing yield risks in the United States and Japan are beyond human control in the sense that they cannot be reduced or prevented. This is what can be expected for highly developed agriculture where the opportunities for increasing productivity and reducing yield variability are generally exhausted. Under such circumstances assuming risks may prove more economical than preventing them. Both Japan and the United States can afford to subsidise agriculture heavily owing to their high per capita incomes and because only a small fraction of the population is dependent on agriculture, so that the subsidies form a very small proportion of their GNP. Subsidies on a similar scale would be prohibitive for a country like India. Also, in evaluating the subsidies for Japanese agriculture one should not overlook the acute scarcity of land and a long record of heavy agricultural taxation for financing development. In the United States as well as Japan, rapid industrial development in the post-war period has led to a significant rural-urban incomes gap, part of which they seem to be bridging through the subsidisation of agriculture.

The conditions in India are vastly different in respect of each of the above factors. The introduction of crop insurance in India at this stage would be premature even on an experimental scale. India can wait and benefit from the experiments in this regard currently under way in many countries. Even an experimental programme in a large country like India would involve the diversion of scarce financial and administrative resources on a significant scale from the more promising sources of raising agricultural productivity and reducing yield instability.

Prepared in 1970 when the author was a member of the Expert Committee on Crop Insurance constituted by the Government of India.

11

The Single State Zonal System for Foodgrains: Increasing Disparities

The variety in soil-climate conditions in India has led to a substantial measure of regional specialisation in the production of crops. Such a specialisation contributes to the optimum utilisation of natural, human and other resources to the advantage of the country as a whole. But in times of acute shortages of food, such as the current year, regions which specialise in the production of food crops are placed in an unduly favourable position at the expense of areas which specialise in non-food crops and which earn valuable foreign exchange for the nation.

It is, therefore, natural that there should be a widespread demand for an equitable distribution of the available foodgrains between different regions. It has been contended in this connection that the abolition of the present single-State food zones and the restoration of free trade in the inter-State movement of foodgrains would ensure a better distribution and a more equitable incidence of shortages.

Genesis of Single State Zones

It is often argued that the single State zones have been formed in order to facilitate intensive procurement in surplus States and that the inter-State disparities in the per capita net availability of foodgrains have narrowed down as a result of the transfer of foodgrains from surplus to deficit States through such procurement. The exponents of this view also contend that the increasing disparities in the market prices of foodgrains between different States do not reflect the inequitable character of distribution. Such a comparison, they suggest, would be valid only if the weighted prices consisting of controlled and free market prices are used, as a substantial part of the demand in deficit States is met through supplies at controlled prices.

It is true that the cordoning off of surplus pockets would minimise the competition from traders and facilitate procurement by the State. But it is difficult to agree that the present single State zones have actually come into being with such an altruistic purpose. Procurement could also be possible through a system of levy on the cultivators while removing the ban on the free movement of foodgrains among States. Moreover, it is well-known that there are surplus as well as chronic deficit pockets within each State. The Government could procure sizeable stocks by cordoning off these surplus pockets allowing free trade in the rest of the State as well as between adjoining areas of other States. In Andhra Pradesh, for instance, the bulk of the marketed surplus of rice is provided by the Krishna, and East and West Godavari districts. Procurement would perhaps become more efficient if efforts are concentrated in such compact areas.

The truth is that single-State zones were the direct result of the anxiety on the part of surplus States to meet their own consumer needs first. It may be recalled that large zones began to crumble as shortages increased and the surplus States started restricting the free outflow of foodgrains unilaterally. Owing to the shortages, the prices registered a significant rise even in the surplus States, although the rise has been much sharper in the deficit States. Thus the interests of the consumers in the surplus States outweighed the interests of the producers who are getting higher prices anyway. It is easy to imagine that if surpluses become sizeable as a result of a succession of good harvests, the interests of the producers would assert themselves and the surplus States would become powerful advocates of free movement of foodgrains.

Inter-State Disparities in the Availability of Foodgrains

There may be some evidence to suggest that the inter-State disparities in the net availability of foodgrains per head declined in 1964-65 when compared to the previous years. But the position in regard to total cereals is not a satisfactory guide. The report of the Agricultural Prices Commission indicates that "the variation in the availability of wheat and of all cereals as between States has declined but the variation in the availability of rice has increased in 1965 in comparison with the variation in 1961-63". It may be noted that the variability of rice increased despite the bumper crop and a record procurement last year. The position must be much worse this year as the shortfalls in output are especially significant in regard to rice. Evidently, the variability in the total supply of cereals has declined owing largely to the increased imports of wheat and their use in the wheat-eating deficit zones. This is by no means a true measure of success in regard to internal efforts and the credit for it can hardly go to the single-State zones.

It is true that free market prices cannot provide the real measure of burden when substantial quantities of foodgrains are supplied at controlled prices. But even if the weighted prices indicate a decline in their variation between different States, it does not follow that the foodgrains are being distributed equitably. When free inter-State movement is prohibited and specified quantities are transferred and distributed by the Government on controlled prices inequitable distribution need not be reflected in the widening of disparities in weighted prices. On the other hand, the disparities in the free market prices beyond the transfer costs do indicate that more grain would flow from surplus to deficit States if restrictions on free movement are removed. Indeed, if Government procurement in the surplus States is intensive and transfers to deficit States liberal, it would have the effect of raising the free market prices in the former and lowering them in the latter, thus narrowing down the gap between the two.

Whatever the system of inter-State distribution, it is recognised that each State — whether surplus or deficit — has a responsibility for procuring adequate quantities of foodgrains internally and making them available to the vulnerable regions and income groups within the State at reasonable price. This is not merely desirable but inevitable in a democratic system wedded to the goals of equity. So long as this right is reserved for each State, the mere abolition of single-State zone system and the restoration of the free movement of foodgrains may not ensure an equitable distribution in periods of acute shortage. The surplus States can always lay their hands on sizeable marketed surpluses for the purposes of local consumption.

Thus even with the free movement of foodgrains between States or alternatively with the active role of the Central Government in the procurement and distribution, it would always be possible for the surplus States to consume significantly more than what is warranted by the overall scarcity condition. In such a case the narrowing down of the disparities in free market prices between different States as a result of free movement would be no indication of an equitable distribution.

Equitable Sharing of Distress

The crux of the problem, therefore, is to make the surplus States agree to share the distress equitably with the rest of the nation. This has to be accomplished through a combination of persuasion and sanctions by the Central Government. Apart from the anxiety to give preference to the local consumption needs, the surplus States have developed a vested interest in holding the stocks of foodgrains. Food is being used as a bargaining counter in their dealings with the Central Government as well as the deficit States. These States might continue to use food as a political weapon in a variety of subtle ways whatever be the system of inter-State distribution. The deficit States, on the other hand, do not have a direct command over the fruits of their specialisation and are, therefore, unequally placed in this process of bargaining. The Central Government has, therefore, to seize upon the necessary counter-veiling measures on behalf of these States in dealing with the surplus States.

The single-State zone system makes it easier for the surplus States to bottle up the desired quantities for local consumption than under any other system of transfers between States. This explains the increasing disparities in the foodgrain prices between surplus and deficit States. The Report of the Agricultural Prices Commission gives an illuminating comparison of wholesale prices of rice between 1960-61 and 1964-65: "The West Bengal price which used to be 2 per cent below the Andhra price is now 14 per cent higher. The Kerala price which used to be 6 per cent above the Andhra Pradesh price is 11 per cent higher inspite of the fact that the 1964-65 index is based on control prices. And the Gujarat price which used to be 1 per cent below the Andhra Pradesh price is 21 per cent higher". These differences persisted despite last year's bumper crops. It is not surprising, therefore, that the price disparities should have assumed fantastic proportions during the current year suggesting that the consumption in the surplus States is being subsidised at the expense of the consumers of the deficit States. Experience has shown that in the period of shortages it is extremely difficult for popular governments to procure and transfer from one State to another as much as the Free Trade would do. Continuation of the single-State zone system, therefore, imposes severe hardships on the people of deficit States and undermines considerably the sense of national unity. The abolition of these zones is, therefore, an economic as well as a moral imperative.

Along with the restoration of free trade in the movement of foodgrains between different States, intensive efforts should be made for building up stocks with the Central Government. This may be done in competition with the traders, if possible, or by cordoning off the most promising surplus districts, if necessary. These stocks should be used for providing a feasible minimum to the vulnerable regions and income groups at reasonable prices. The existing stocks with the Central and State Governments and the likely increases in them do not justify the promise of rationed distribution in all the towns with a population of one lakh and above. At any rate, the injustices associated with the present zonal system should not be perpetuated by postponing the question until rationing is introduced in these towns, which may be beyond the realm of possibility during the current year.

The Times of India, February 18, 1966.

12

On Indian Agrarian Scene

Indradeep Sinha's book* appears at an opportune moment, as the attention is now focussed on the agrarian scene in India, thanks to the farmers' agitation for 'remunerative prices' for agricultural produce. The publication is the revised version of the general secretary's report presented by the author to the 22nd National Conference of the All-India Kisan Sabha held at Vijayawada in June, 1979. The work is both comprehensive and authoritative and hence merits serious study by all those interested in an understanding of the emerging agrarian relations in India as well as the prospects and strategies for change.

Characterisation of the Existing Agrarian Relations

From the data and the analysis presented in Chapter Six on 'Changes in the Agrarian Set-up', the author arrives at a reasonable generalisation that the present agrarian economy of India is a "multi-structural economy consisting of the capitalist, the small peasant and the semi-feudal sectors coexisting but also contending with each other"[1] (p. 138).

The characterisation made by the author admits of regional diversity as any such attempt ought to, considering the vastness of the country with its continental dimensions. In a place like North Bihar, for instance, semi-feudal relations may be breaking up slowly whereas in a region like Punjab, capitalist relations

* "The Changing Agrarian Scene: Problems and Tasks", Peoples Publishing House, 1980.

may be the dominant characteristic. However, the generalisation that, for the country as a whole, "the capitalist sector has already become the leading sector of Indian agriculture" appears to overstate the position, considering the facts brought out by the author.

During the 18 years between 1953-54 and 1971-72, the author points out, "both the top groups — landlords (above 25 acres) and rich peasants (10–24.99 acres) — lost nearly 39.5 million acres of land, while the three lower groups — middle peasants (5–9.99 acres), small and poor peasants (1–4.99 acres) and semi-proletarian agricultural labourers (below 1 acre) together gained nearly 25.4 million acres of land.... The area owned by landlords (above 25 acres) declined by nearly 24 million acres due to real or fake partitions and *benami* transfers with a view to evade ceilings or due to sale to well-to-do peasants of the lower groups, particularly in areas of militant land struggles, or due to surrender to the government under the ceiling laws (2 million acres) or through transfer to tenants and sharecroppers as a result of their prolonged militant struggles" (p. 119).

The contribution of tenancy reforms and ceiling on the ownership of land consisted not so much in the acquisition of 'surplus' land by the State from the landlords and their distribution among the landless and the sharecroppers as in arresting or slowing down the development of capitalist farming through the expansion of owned area. There is thus a serious constraint on the development of capitalism in agriculture because of the ceiling on the ownership of land, although ownership of land is not essential for the development of capitalist agriculture. So long as there is no ceiling on the operational holding, a capitalist farmer can lease in area and augment his operational holding.

However, according to the data analysed by the author, the area operated (owned land operated plus land leased in) by holdings of 10 acres and above, declined from 65.84 per cent of total operated area in 1953-54 to 53.23 per cent in 1971-72 (p. 120). As pointed out by the author, this decline is explained not only by the *benami* transactions by the large landowners but also by subdivisions, sale of land under the impact of ceiling legislation, tenancy reforms and acquisition of surplus land under the ceiling laws.[2]

In Punjab, households above 10 acres "while constituting only about 45 per cent households leasing-in land, lease in more than 70 per cent of the total area leased-out.... This tendency towards not cultivating even the owned land has been growing among poor and small peasants in Punjab during the last 25 years" (pp. 122-23). Whereas the author recognises that this situation "explains the absence of a tenants' or sharecroppers' movement in Punjab although it leases-in a larger percentage of the operated area than West Bengal", he does not spell out the implications of this pattern of capitalist development for the existing legislation on 'security of tenure' and 'fair rents', which can benefit essentially the capitalist farmers in areas like Punjab. Also there is no discussion of the desirability of ceiling on operational holding for arresting the growth of capitalist farming at the expense of the small and marginal farmers.[3]

If non-land assets are also taken into account there is an evidence of an increase in inequality in the ownership of total assets by rural households between 1961-62 and 1971-72. The author quotes the Reserve Bank to show that "the poorest 10 per cent of rural households owned only 0.1 per cent and the richest 10 per cent owned more than half of the total assets in 1971-72 as well as 1961-62" (p. 48). The author points out that "in the absence of equitable distribution of land, even the development of irrigation, similar other inputs and technical extension services, serve only to strengthen the monopoly position of the big landlords", and that "this group already enjoys a monopoly position in the utilisation of institutional credit as well as in command over the bulk of agricultural products sold in the market" (p. 68).

In view of these inequalities in the ownership of assets and access to resources, the author suggests that "reliance on technological improvement (*e.g.* through the 'green revolution') on the basis of existing property relations, was one of the primary causes of the growth of unemployment, poverty and social inequality during the plan period in the rural areas" (p. 59). Citing the Tamil Nadu experience, the author states that "it has underlined the inherently anti-peasant character of capitalist development of agriculture. It has indicated that violent conflicts and massive collision are to be expected in state after state in future, if development of agriculture along the capitalist path is vigorously pushed forward on the pattern of Tamil Nadu" (pp. 109-10).

The above quotations are drawn to show that the agricultural policies of the government are heavily biased in favour of rich farmers who reveal strong capitalist tendencies.

Tasks Before the Peasant Movement

In view of the emergence of capitalist farming, the author states that the slogan of "abolition of landlordism and land to the tiller" is no longer adequate and that there is need to add two more fundamental slogans, *viz*., "Remunerative prices for agricultural produce" and "Preferential aids to the small peasantry for developing agricultural production" (p. 152). However, a full reading of the book makes it abundantly clear that the focus of the "tactical line of a diversified mass movement" is on "remunerative prices for agricultural produce, so as to fully cover the cost of production, including a fair remuneration for the peasants' labour; and for full parity between the prices of agricultural and industrial products, between the prices of inputs and outputs of agriculture and between the prices paid and the prices received by the peasantry" (p. 152).

With the emphasis given to 'remunerative prices' as a rallying point for the peasant movement, the book presents two sharply contrasting and mutually contradictory positions: The characterisation of agrarian relations as predominantly capitalist where a small minority of big farmers account for bulk of the agricultural resources and benefits; and the united front of all classes of farmers including capitalist farmers for 'remunerative prices', 'parity prices' *etc*. from which the greatest beneficiaries would be the capitalist farmers.

The author pleads for forging "widest possible unity of action in every struggle and movement without any political inhibitions" (p. 161), through the formation of the separate multiclass organisations "like the organisations of cane growers, coconut growers, rubber growers *etc*." (p. 154).

Despite the characterisation of existing agrarian relations as predominantly capitalist, the question of minimum or fair wages to agricultural labour does not form an element in the main planks of the mass movement. Indeed, it is mentioned in the preface itself that "problems relating to the development of the agricultural labourers' movement have been omitted as they fall outside the purview of a Kisan Sabha report". This may be all right in a formal sense but its omission appears glaring in a report which calls for the widest possible mass movement for improving the condition of the rural poor.

On the question of fair prices for farm produce, it is necessary to examine in detail (a) whether the prices of farm produce cover the cost of production and leave a fair margin of profit; (b) whether the prices received by the farmer for his produce have been lagging behind the prices paid by him for inputs and other non-agricultural consumption goods; and (c) who are the beneficiaries and the losers from the observed behaviour of agricultural prices in relation to cost of production and prices of non-farm goods. These are, essentially, questions of fact and it should be possible to form rational and correct judgments on these issues by appealing to the necessary data and studies on the subject.

Cost of Production and Prices of Farm Produce

The Agricultural Prices Commission while recommending prices for agricultural commodities takes into account the cost of production estimated under the "comprehensive scheme" of the Government of India. These cost estimates include, among other elements of cost, the wages for hired labour and the value of family labour imputed on the basis of the actual wage rates prevailing for the hired labour. The main point of difference raised by the author is that human labour input should have been valued on the basis of the minimum wages fixed by the Government instead of the prevailing market wages:

> "According to the data supplied by the Agricultural Department, the per quintal cost of production of paddy in West Bengal in 1977-78 was Rs. 70.93. Out of this, "cash and kind expenditure" plus the value of "family labour" amounted to Rs. 41.03 per quintal. The value of "family labour", as we have seen in the case of jute, is calculated at "the prevailing rates of wages" which are only 52 per cent of the minimum wages fixed by the West Bengal Government. Since human labour forms nearly 30 per cent of the cost of production (and nearly 50 per cent of cash and kind expenses, *etc.*) the actual cost would work out at Rs. 91-92 per quintal in West Bengal in 1977-78 which exceeds the minimum price (Rs. 85.00) by Rs. 6-7 per quintal". (p. 90).

In is clear from the above that the support price of paddy (Rs. 85.00 per quintal) did cover the cost of production (Rs. 70.93 per quintal) and left a significant margin in 1977-78 when the human labour input is valued on the basis of the 'actual market wages' but it did not cover the cost amounting to Rs. 91-92 per quintal when human labour input is valued on the basis of 'minimum wages' fixed by the Government. It is important to note that under the procedure followed for computing the cost of production, the minimum wages are automatically taken into account whenever they are enforced because human labour is valued according to the actual wages paid. It is also important to note the consequences of valuing human labour on the basis of minimum wages, even when the latter are not in force. The capitalist farmers and others who hire labour appropriate surplus to the extent of the difference between the market wages paid and the legally fixed minimum wages.

Secondly, the price of the produce — in this case paddy — is raised for the wage earners without raising their wages. The agricultural labour is at a loss on both counts — by not getting the higher (minimum) wages and by having to pay higher price for food. The author should have stressed the need to ensure minimum wages in practice rather than advocating the escalation of prices on account of minimum wages. To the extent these wages are enforced in practice, they would automatically be reflected in the cost of production estimates.

If human labour is to be valued on the basis of 'minimum wages' fixed by the Government, then one can argue by the same logic, that the services of land should be valued on the basis of 'fair rent' fixed by the Government instead of the present practice of valuing it on the basis of market rent. Since 'fair rents' are significantly lower than the actual market rents and since rent forms a sizeable component of cost of production — larger than wages in most parts of the country — valuation of all the land on the basis of 'fair rent' would underestimate the cost of production, even when human labour is valued on the basis of 'minimum wages'. However, to the extent 'fair rents' are implemented, they would be reflected in the cost of production estimates.

In a detailed analysis of the evidence on cost of production and procurement prices for wheat, covering the period up to late sixties, G.S. Raychoudhury and Raj Krishna conclude that "the wheat procurement price did not cover the full cost of production, (or exceeded it only by a small margin in some irrigated tracts) in the fifties, but it provided a comfortable margin above full cost in the early years of the green revolution. In contrast with the procurement price, the farm harvest price, namely, the price prevailing during six to eight weeks immediately after the harvest in each State, has been above the cost of production in all important wheat producing states in the fifties as well as sixties".[4] The proportion of wheat procured by the Government to the total output of wheat varied between mere one per cent to four per cent during the period 1959-60 to 1966-67 and rose steadily thereafter reaching a figure of 23.3 per cent in 1976-77.[5] It is thus clear that in the early period when procurement prices did not cover the cost of production, bulk of the produce was in fact sold at market prices which covered the cost of production and, in the later period when public procurement became substantial, both procurement and market prices were above the cost of production.

In a similar study for rice over the same period, they conclude that, "as in the case of wheat, the farm harvest price of rice has been above the cost of production in most of the states. But in Andhra Pradesh, parts of Bihar, West Bengal and Uttar Pradesh, even the farm harvest price has been less than the full cost of production.... the rice procurement price has been below the cost of production in West Bengal and South India, though in recent years the gap has tended to shrink. In the northern states the procurement price has usually covered the full cost of production."[6]

Again, it may be noted that the procurement of rice by the Government constituted only about 1 to 2 per cent of total output of rice in the country for the decade ending mid-sixties and steadily rose to 12 per cent in 1975-76.[7] In West Bengal, this ratio has been well within 5 per cent throughout the period ending 1975-76; the maximum ratio reached is around 20 per cent as against about 75 per cent to 80 per cent in Punjab and Haryana.[8] Whereas the picture for rice is mixed and is not as encouraging as for wheat, it may, nevertheless, be noted that the farm harvest prices generally covered the cost of production and procurement of rice has been substantial precisely in those states and periods where the procurement prices covered the cost of production.

In a recent study covering the period between 1967-68 and 1976-77, K. Subbarao shows that "in the case of wheat in Punjab the procurement price was nearly always as high as the open market price and covered the cost of production throughout the period".[9] His study also indicates that procurement price for rice in Andhra Pradesh which "was considerably lower than the market price throughout the period,"[10] may not have covered the cost of production up to the early seventies, but seems to be well above the cost of production since the mid-seventies.

Relationship Between the Prices Received and Prices Paid by the Farmers

Another major contention in the book is that the terms of trade have been turned against agriculture. The author arrives at this conclusion by comparing the index numbers of wholesale prices of manufactured products and agricultural products for the period 1971-72 to 1976-77. He computes the loss suffered by the agricultural sector which approximates between Rs. 900 crore in 1971-72 to 1500 crore in 1975-76 (pp. 29-30).

It is necessary at this stage to draw attention to certain methodological issues involved in the study of terms of trade. In view of the weather-induced fluctuations in agricultural output and prices, computation of losses and gains on the basis of any single selected base year becomes highly hazardous. In fact, the figures would change drastically by changing the base years. The author shows, for instance, that "between 1974-75 and 1975-76, although the quantum of production increased by about 11 per cent but prices fell by about 16 per cent" (p. 30). This arises from the inelastic nature of demand for agricultural commodities. For the same reason, the prices can be expected to increase more than proportionately for a given decline in agricultural production. Secondly, on account of this variability in agricultural output, one should take a fairly long period for analysis and study the trends over a period rather than the figures for particular individual years in relation to some selected base year. Thirdly, the comparison with the price index of the manufactured products as a whole, as done by the author, is not quite appropriate. Instead, comparison should be made with the prices of only those manufactured products which enter into the cost of production for agriculture and into the consumption basket of the agricultural population, with appropriate weights assigned to each one of them on the basis of the cost of production data as well as consumption surveys. Lastly, the prices used should be the actual ones at which farmers buy and sell products.

R. Thamarajakshi has done a pioneering study on this subject by using appropriate methodology covering 24 years from 1951-52 to 1974-75. Her study, used extensively by professional economists, concludes as follows:

> "It will be observed that during this period while all prices received by agriculture rose at 5.94 per cent per annum, those paid by agriculture registered an annual rate of increase of 4.45 per cent; consequently, the net barter terms of trade improved in favour of agriculture at 1.43 per cent per annum. It is further noteworthy that since 1964-65, the net barter terms of trade have been consistently favourable to agriculture, the average level of the index being over 120 during the eleven-year period ending 1974-75. Moreover, the rate of improvement in the terms of trade of foodgrains compared to the non-agricultural commodities purchased by agriculture for all uses is equal to that in the overall

terms of which all commodities have been exchanged between the two sectors during this period. What is even more striking is that the prices received by agriculture have risen at a faster rate (5.94 per cent) than those of intermediate products purchased by agriculture from non-agriculture (4.53 per cent). As a result, the terms of trade of all agricultural products sold to non-agriculture *vis-a-vis* the non-agricultural intermediate products purchased by agriculture have improved in favour of agriculture at a compound rate of 1.34 per cent. Compared to this, the prices paid by non-agriculture for agricultural raw materials have recorded a faster rate of increase (5.89 per cent) than those paid by agriculture for all non-agricultural products (4.45 per cent). Finally, the index of income terms of trade which is a true reflector of the economic betterment of agricultural sector has grown at as high a rate as 4.53 per cent per annum during this period thereby showing the increasing capacity of this sector to purchase non-agricultural commodities. Thus, looked at from any angle, the terms at which trade has been transacted between agriculture and non-agriculture have been improving in favour of the former, since 1951-52".[11]

Agreeing with the analysis done earlier by Thamarajakshi for the period 1952-53 to 1962-66,[12] Ashok Mitra, former Chairman of the Agricultural Prices Commission, observes: "Whatever the contributory factors, the general inflationary trend has been overshadowed, in the course of the past decade, by the phenomenon of agricultural prices gaining perceptibly over industrial prices".[13]

However, the idea of comparing the fortunes of 'agricultural sector', as a whole, cutting across all classes, with that of 'non-agricultural sector' is itself open to question.[14] In particular, the concept of parity prices has been seriously questioned by economists on grounds of its adverse impact on efficiency as well as the distribution of income. The concept of parity price has its origin in the growth of capitalist farming. Since capitalist sector in agriculture is market-oriented for the sale of output as well as for the purchase of inputs and consumption goods, a decline in agricultural prices relative to those of inputs and consumption goods would affect it adversely. Small and medium farms, on the other hand, are relatively less affected because they use family labour on the farm and produce essentially for self consumption.

The concept of parity prices was first used in the United States of America in 1921 by the president of a tractor company fearing that the reduced income of farmers would adversely affect their sales of tractors. The experience in the Soviet Union and other socialist countries has been different. In these countries, despite the absence of the capitalist sector in agriculture, terms of trade were frequently shifted against agriculture, to mobilise resources for industrialisation. The Special Expert Committee on Cost of Production Estimates headed by S.R. Sen after examining in detail the desirability of constructing a parity index of prices received and prices paid by farmers concluded as follows:

"The Committee does not believe that the parity price approach will necessarily help advance income parity. The well documented US experience provides ample evidence to prove that such a policy has helped mainly the large farmers who have large surpluses to sell. Large farmers buy most of the subsidised inputs and sell most of the price supported output. Small farmers use fewer purchased inputs and sell a small amount of output.

"Parity-based price support escalating costs of inputs, specially the land value, has been extensively reported. Although, all land owners receive the benefit of capital gain, the large farmer, under expectation of higher output prices, is best able to bid higher prices for additional land. Land values rising disproportionately produce a ratchet effect on costs and are generally disadvantageous to the small farmer and the landless.

"The objective of parity income, therefore, can best be achieved through structural changes in land-ownership, public investment in irrigation, distributing subsidised inputs to small farmers, *etc*. Price mechanism, relatively a weak instrument in itself for the purpose in an imperfect economy, may contradict the other welfare programmes if the parity price approach is adopted mechanically."[15]

Agricultural Prices and Incomes

As mentioned earlier, the impact of agricultural prices on the incomes of the farmers depends on the distribution of marketable surplus among different size groups of holdings. In a careful study, Utsa Patnaik found that in 1960-61, for the country as a whole, "The importance of 'small' holdings below ten acres — which make up some 82 per cent of holdings and account for 75 per cent of the cultivating population excluding labourers — in contributing to total marketable surplus is considerably less than the importance of the "medium and large' holdings above ten acres". She concluded that "roughly about 18 per cent of cultivating households with 25 per cent of population, which concentrate 61.2 per cent of the total operated area, generates over half the gross output from agriculture (52 per cent) and accounts for over two-thirds of aggregate marketable surplus (66.8 per cent)".[16]

There is enough evidence to suggest that this inequality in the distribution of marketable surplus has been accentuated in the last two decades because of the introduction of new technology and its greater adoption among the larger farms who have better access to resources.[17] For instance, a recent study in Chingalput district of Tamil Nadu covering the period 1970-75 shows that the small size group (below five acres) including tenants contributed just 8 per cent to the total marketed surplus, the middle size group (5-10 acres) contributed 20 per cent; and the large holdings (above ten acres) contributed the remaining 72 per cent of the surplus.

As the author rightly observes, the poor and middle peasants are "generally compelled to sell at comparatively low prices at the time of the harvest and purchase the same or other foodgrains during the lean months at twice the harvest prices or even more" (p.14).[18] It is, therefore, necessary to consider the distribution of net marketed surplus (*i.e.*, net of repurchases) to know the impact of prices on the distribution of income.

In a study covering 15 villages and 150 cultivating households in Hoogly district of West Bengal for the period 1970-71 to 1972-73, Ashoke Hati shows that the net marked surplus was negative up to 0.66 hectares, then remained at about 5 per cent of the net receipts up to 1.98 hectares, and increased thereafter with the increase in the size of holdings.[19]

Another study covering North Bihar by S.P. Sinha, B.N. Verma and D.K. Sinha showed that nearly 80 per cent of the households in the size group below five acres had no surplus to sell at all.[20]

Analysing the relevant data and the available studies on this subject, V.K.R.V. Rao observes: "It is clear, therefore, that a great majority of rural population is not benefited by a rise in foodgrain prices while a substantial portion of the rural population is actually adversely affected by such a rise".[21]

It is clear that under the existing structure of land and asset distribution, farm prices that are higher than the cost of production would result in disproportionately larger benefits to the capitalist sector in agriculture. In the case of foodgrains in particular, where bulk of the procurement and distribution is undertaken by the government, higher procurement prices would in fact redistribute income from the vast majority of the poor both in rural and urban areas — who spend over 80 per cent of their meagre income on foodgrains — to the capitalist farmers. There is now a subsidy of about Rs. 600 crore per annum in the procurement and distribution of foodgrains undertaken by the government. To the extent this subsidy is attributable to the higher procurement prices, it represents transfer of income from the community as a whole to the farmers, at least two-thirds of which accrues to the rich farmers.

Conclusion

If the characterisation of the existing agrarian relations as predominantly capitalist is correct, and if capitalist development in agriculture goes against the interests of the large majority of rural population consisting of small and marginal farmers and landless labourers, then a strategy which focusses on securing higher procurement prices even when the terms of trade have been, by and large, favourable to agriculture for over a decade, would be inconsistent with the above characterisation of agrarian relations, as, such a strategy would contribute to consolidating and strengthening further the capitalist sector in agriculture. Instead, greater priority should be assigned for the second goal set in the book, *viz.*, "increasing agricultural production by improving the husbandry of the poor, small and middle peasants through a policy of preferential aids, including credit and other inputs, to them" (p. 152)[22] while stepping-up the movement for the third goal, *viz.*, "radical land reforms and land to the tiller" (p. 152).

Even in regard to remunerative prices, there is considerable scope for increasing the incomes of small farmers by ensuring that they in fact get the prices fixed by the Government. As the author rightly observes "middle, small and marginal farmers are seldom able to sell their produce at the official support or procurement prices. Similarly, they are seldom able to buy their essential consumer goods and agricultural inputs at the official or ex-mill wholesale prices" (p. 31). In the case of credit and inputs, there is every justification for preferential allocation to the small farm sector at prices significantly lower than those charged to the large farm sector. This can be a potent source for redistribution of income and wealth in favour of the poor.

Notes

1. This reviewer had made the following comments while reviewing the book *Evolution of Agrarian Relations in India* by Bhowani Sen: "These difficulties have inevitably led the author to come out with the conclusion that 'despite changes and modifications in the course of the last decade, Indian agriculture is as yet largely enmeshed in semi-feudal relations... The whole evidence produced in the book, on the other hand, leads one to a simple and commonplace conclusion that the present agrarian relations are characterised by the coexistence of self-cultivating peasants who also hire labour occasionally and land-owners who operate mainly with hired labour. If one likes one can define this relationship as capitalist and semi-capitalist operating at a very low level of production technique" (Seminar 38, October 1962). There is no doubt that the production techniques in agriculture have improved considerably over the last two decades. Under these conditions, owner-cultivation with hired labour has resulted in the development of a rich farm sector or capitalist sector which wields considerable influence on policy-making for agriculture.

2. The following is the overall picture emerging from the various studies based on the available data on landholdings since independence:

   "The inequality in the distribution of land owned in rural India has shown some decline, particularly in the sixties. The inequality in the distribution of area operated has also declined, but the decline is less marked than in the case of owned land. At any point of time, the distribution of operated area happens to be less unequal than owned area, owing to the fact that small and medium holdings lease in area, the bulk of which is leased out by the large landowners. However, there is a tendency towards convergence of the distributions of owned and operated area because the area under tenancy has declined owing largely to the resumption of land by the landowners from the small and medium size operational holdings. Subdivision of land holdings and sale and purchase of land seem to have been more important in bringing about these changes in the structural distribution than the redistribution of land through tenancy reforms and legal ceilings on land holdings. As a result of these changes the number of marginal and small farmers as well as the area held by them has increased, whereas the importance of large farms has declined both in terms of their number as well as the area held." (C.H. Hanumantha Rao, "Rapporteur's Report on Changes in the Structural Distribution of Land Ownership and Use Since Independence," *Indian Journal of Agricultural Economics*, Conference Number, July-September, 1976).

3. For a discussion of the possible impact of ceiling on operational holding on output and employment see C.H. Hanumantha Rao, "Ceiling on Agricultural Land-holding — Its Economic Rationale," *Economic and Political Weekly*, Review of Agriculture, June 24, 1972.

4. G.S. Raychaudhury and Raj Krishna, "Some Aspects of Wheat Price Policy in India," *Indian Economic Review*, Oct., 1979.

5. *Ibid.*

6. G.S. Raychaudhury and Raj Krishna, "Some Aspects of Wheat and Rice Policy in India," *Indian Economic Review*, April, 1980.

7. *Ibid.*

8. *Ibid.*

9. K. Subbarao, "Producer Levy, Evasion and Income Loss: Empirical Evidence from Coastal District of Andhra Pradesh," *Economic and Political Weekly*, Review of Agriculture, March 24-31, 1979.

10. *Ibid.*

According to the report of the Agricultural Prices Commission on *Price Policy for Wheat for the 1979-80 Crop* the total cost of production per quintal for wheat (including the rental value of own land and imputed value of family labour) in 1977-78 in the two major wheat growing states *viz.*, Punjab and Haryana was Rs. 108.57 and Rs. 104.01 respectively. (The corresponding figures for 1978-79 are 101.45 and 114.00) as against the recommended minimum price at Rs. 117.00 per quintal for 1979-80.

The procurement price for paddy for the 1979-80 marketing season was fixed at Rs. 95 per quintal for 'common' varieties, as against the total cost of production per quintal of Rs. 88.36 in Andhra Pradesh and Rs. 67.53 in Punjab for 1978-79.

For the same year, the government fixed a minimum price of Rs. 12.50 per quintal for sugarcane (for a basic recovery of 8.5 per cent) as against the total cost of production (as defined above) per quintal of Rs. 7.99 and Rs. 8.50 in Uttar Pradesh and Bihar respectively for 1976-77 and Rs. 10.98 and 7.62 per quintal in Andhra Pradesh and Maharashtra respectively for 1977-78. The present demand for price of Rs. 30 per quintal can be justified only under the assumption that the cost of production has tripled in the last three years.

11. R. Thamarajaskshi, "Role of Price Incentives in Stimulating Agricultural Production in a Developing Economy,' in *Food Enough or Starvation for Millions*, edited by Douglas Ensminger, Tata McGraw Hill, New Delhi, 1977.

12. R. Thamarajkashi, "Intersectoral Terms of Trade and Marketed Surplus of Agricultural Produce, 1951-52 to 1965-66," *Economic and Political Weekly*, Review of Agriculture, June 1969.

13. Ashok Mitra, *Terms of Trade and Class Relations — An Essay in Political Economy*, Frank Cass, 1977, p. 109.

He points to the "Government's discriminatory attitude with respect to the fixing of administered prices *apropos* of wheat and rice on the one hand, and cotton and jute on the other" and offers the following explanation:

'In the case of jute — and, to a large extent, rice — the majority of cultivators are mostly small farmers and share-croppers, who are weak, unorganised, and have but little influence on the processes of the polity. In contrast, the bulk of the wheat crop — and certainly the overwhelming proportion of its marketed surplus — are raised in relatively large-sized holdings. The cultivation of cotton too is increasingly being organised on commercial and quasi-commercial sales in big holdings; besides, the relatively affluent amongst the farmers who raise the cotton crop have intimate links with trade and industry and are accordingly in a position to bring a strong influence to bear upon the authorities in the matter of pricing policy.' (*Ibid.*, pp. 130-31).

14. *Ibid.*, pp. 44-68.

15. Ministry of Agriculture, Government of India, *Report of the Special Expert Committee on Cost of Production Estimates, 1980.*

16. Utsa Patnaik, "Output and Marketable Surplus of Agricultural Products: Contribution by Cultivating Groups, 1960-61," *Economic and Political Weekly*, Review of Agriculture, December 27, 1975.

17. C.H. Hanumantha Rao, *Technological Change and Distribution of Gains in Indian Agriculture,* Macmillan, Delhi, 1975; reprinted in 1980; Chapter 11.

18. S.S. Sivakumar, "Aspects of Agrarian Economy in Tamil Nadu: A Study of Two Villages, II, Constraints on Production and Marketing," *Economic and Political Weekly*, May 13, 1978.
19. Ashoke Hati, "Non-linear Marketable Surplus Functions," *Economic and Political Weekly*, July 17, 1976.
20. S.P. Sinha, B.N. Verma and D.K. Sinha, "A Study of Marketed Surplus, Scale Elasticity of Surplus and Prospect of Mobilising Surplus in North Bihar," *Indian Journal of Agricultural Economics*, July-September, 1975.
21. V.K.R.V. Rao, "Does Rural India Benefit by Rising Foodgrain Prices?" *Commerce*, July 28, 1979.
22. Stating that 'An improvement in the level of farm prices by itself would obviously not lead to an improvement in the economic conditions of the majority of the farming population,' Ashok Mitra suggests that 'If adequate incentives need to be offered to the middle peasants, such incentives could assume the form of special dispensations for the purchase of inputs, including fertiliser, seeds and irrigation water, and in the supply of credit.' See Ashok Mitra, 'Wage-Labour and Wage-Goods: Some Policy Issues' *A.K. Gopalan Memorial Lecture-II*, Indian School of Social Sciences, Trivandrum, 9 April, 1980.

Link, January 26, 1981.

13

India's 'Surplus' Cattle: Some Empirical Results

The recent debate on India's "surplus" cattle has revealed that the area of controversy has considerably narrowed down. For instance, V.M. Dandekar who believes that India's present cattle number can be reduced profitably has stated that "India needs all the draught animals that there are; with the existing technology and institutional arrangements, their number cannot be reduced."[1] In the case of cows, however, he observes that "a smaller number of better-fed cows will contribute more milk and dung and will give enough calves to replace the needed stock of draught animals."[2] This observation is based upon his view of the nature of the production function underlying the Indian cattle economy, *viz.*, "with better feeding their productivity can be more than proportionately increased."[3] K.N. Raj who has questioned the widespread assumption about the existence of large surpluses of livestock in India raises the similar question, *viz.*, "whether, given the amount of food that can be spared for maintaining animals in the situation, a higher output is secured by giving this food to a smaller number and raising their productivity or by distributing it over a larger number" and suggests that "a firm answer to this question cannot be offered without much deeper investigation into the relevant production functions."[4]

There is hardly any empirical evidence bearing on this problem. There is no doubt, however, that this question can be settled only by investigating the nature of the production functions underlying the Indian cattle economy. This note is an attempt in this direction.

If a significant proportion of the existing female cattle population is surplus or redundant then one should expect the elasticity of output (*i.e.*, breeding efficiency, milk and dung) response for an animal with respect to the feed and other inputs to be more than unity. Elasticities significantly less than unity could mean that the existing number is less than the number desired by the farmers on economic grounds. The divergence between the desired or the optimal and the actual stock can arise due to factors beyond the control of the farmers. In the case of the elasticity being more than unity, the divergence could be due to prohibition of cow slaughter. In the second case, the divergence may arise owing to epidemics and diseases which farmers are not in a position to control.

If a farmer is found to be getting rid of his animal through deliberate starvation when cow slaughter is prohibited, it may be either because the animal is surplus or is otherwise disease-ridden and useless, even though the actual stock is less than the desired stock. The fact that farmers do resort to deliberate starvation for getting rid of redundant animals leads one to expect the actual stock not to be higher than the desired stock. The actual stock could, however, be lower than the desired stock owing to the incidence of epidemics and diseases. Dandekar suggests that in either case the slaughter of redundant animals may be socially more efficient than killing through starvation as it would augment the food resources of the community.[5] But legalisation of cow slaughter need not necessarily be Pareto-wise optimal as is sometimes believed,[6] because when a promising market for beef is opened, the new equilibrium prices of milk and draught animals could be higher as a result of the slaughter of animals on a large scale. If this possibility is accepted, then legalisation of cow slaughter throughout the country may not prove to be an efficient method until mechanisation of farm operations reduces the demand for draught animals and improvement in farm productivity releases more area from crop production thus raising the profitability of livestock enterprise.

The cow is maintained in rural India mainly for its breeding or bullock-producing function. The cow population constitutes about 75 per cent of bullock population.[7] Dandekar has argued that this ratio can be reduced profitably to 40 per cent by reducing the mortality among the youngstock to half its present level (at present only about 35 per cent of them survive the first three years of life), which needs no more than a little better care and feeding of the youngstock.[8] It is, therefore, necessary to investigate whether, with the kind of feed that the farmers are able to allocate for cows and calves, the breeding efficiency of cows and the survival rate of youngstock can be raised by increasing the feed intake per animal.

Data on this aspect which could be used for a systematic analysis are not readily available. However, the Seventeenth Round (1961-62) of the National Sample Survey throws some light on this aspect: Farmers holding below ten acres own about two-thirds of the total stock of cattle accounting for only about 40 per cent of operated area and 19 per cent of pastures and grazing land. Thus the remaining (large) farmers who account for as much as 60 per cent of operated area and 81 per cent of pastures and grazing land maintain only about one-third of the total cattle stock (Table 13.1). There is, therefore, reason to believe that the per head feed intake of the female and youngstock increases with the increase in the size of operational holding. On this basis one should expect the ratio of youngstock to female stock to rise with the increase in the size of holding if better feeding results in a cow dropping, on an average, more calves and in raising the survival ratio of such calves.

Table 13.1

Percentage Distributions of Number of Cattle, Operated Area and Pastures and Grazing Land According to the Size of Operational Holding: All-India (All Households), 1961-62

Size of Operational Holding (Acres)	*Cattle*				*Lands*	
	Males Over 3 Years	*Females Over 3 Years*	*Youngstock*	*Total Cattle*	*Operated Area*	*Pastures and Grazing Land*
	(1)	*(2)*	*(3)*	*(4)*	*(5)*	*(6)*
Upto–4.99	42.28	43.62	44.35	43.19	19.18 (62.88)	5.26
5.00–9.99	24.97	23.39	23.49	24.13	20.70 (19.21)	13.88
10.00–19.99	18.21	17.71	17.67	17.93	23.78 (11.56)	22.14
20.00–49.99	11.37	11.76	11.32	11.47	24.74 (5.37)	32.07
50.00 and above	3.17	3.52	3.17	3.28	11.60 (0.98)	26.65
	100.00	**100.00**	**100.00**	**100.00**	**100.00**	**100.00**

Note: Figures in brackets give the percentage distribution of operational holdings.
Source: The National Sample Survey, 17th Round, 1961-62 Land-holding Inquiry (Rural), Number 146, pp. 43, 45, 70 and 120.

However, the percentage distributions of female and youngstock are similar indicating that the ratio of young to female stock does not rise with better feeding possibilities. It is possible that with the quality of feed that the farmers are able to spare for the female and youngstock after meeting the feed requirements for bullocks, the increased feed intake per head does not have a significant effect on breeding efficiency. It is also possible that owing to the adequate supply of family labour, small farmers are in a position to take better care of their animals than the large farmers who have to hire labour. The above evidence suggests that it may not be possible to increase breeding efficiency by reducing the female stock without any regard for the quality of available feed and costs of hired labour.

It is important to note this because surplus cattle is expected essentially among large farms having three cows or more.[9]

To study the output response of milk and dung to the available resource inputs, we need to have data on inputs and output per milch cattle in rural areas. A rural-urban comparison, for instance, would not be helpful as the input mix differs significantly, especially with respect to the share of concentrates, giving a high output response.[10] Also, experimental data when the animals are stall-fed under confinement may not be quite useful even if the feed used corresponds to the actual farming conditions. For one thing, farmers have to depend on open grazing of their animals for a good part of the year and regular stall-feeding does not correspond to this real world situation. Secondly, animals fed under confinement can be expected to show a higher output response than when they procure the same quantity of feed through open grazing. Open grazing represents the substitution of animal labour for human labour towards procuring feed and a good part of feed is expended to meet this energy requirement. Similarly, cattle dung realised per animal would necessarily be lower under open grazing than under confinement because of outdoor droppings. Therefore, under conditions of open grazing it is of little use for the farmer to know that "more dung is produced when a given quantity of feed is consumed by one animal than when it is shared by two animals".[11]

Data needed for a study of this kind are rare to find. The data on inputs and output (milk and dung) relating to milch cattle available in published form from the Farm Management Studies for Punjab and Andhra[12] are of some use. These data are aggregated according to different size groups of operational holdings. Separate data on inputs of feed and labour are not available for Punjab. Since these inputs are neither perfect substitutes nor perfect complements, separate data on them would yield better estimates of production functions. Such data are available for Andhra.

We have estimated the elasticities of output by fitting the Cobb-Douglas Production Function of the type $Y= \alpha X_1^{b1} X_2^{b2}$ where Y is output, a is constant, X_1 and X_2 are inputs and b_1 and b_2 are the elasticities (Table 13.2). The results indicate that the output response of milk as well as dung is significantly less than unity with respect to the feed input in Andhra and with respect to the total input in Punjab. As can be expected for reasons mentioned above, the elasticities relating to dung are generally lower than the corresponding elasticities for milk. The sum of elasticities for feed and labour relating to Andhra are not significantly different from unity. These results suggest that in the regions and for the period studied, the actual number of milch animals may not have exceeded the desired number and in Andhra the actual number seems to have been close to the desired or optimal number. However, more intensive investigations with better data for different regions of the country are needed before firm conclusions can be drawn.

Table 13.2

Some Production Functions for Cattle Economy

Region and Period	No. of Observations	Dependent Variable (Y_1, Y_2)	Elasticities with Respect to Feed X_1	Labour X_2	Total Inputs X_3	Sum of Elasticities	R^2
(1)	(2)	(3)	(4)	(5)	(6)	(7)	(8)
Punjab (1954-55 to 1956-57)	15	Milk (Y_1)	—	—	0.59* (0.17)	0.59	0.48
Andhra (1957-58)	15	Milk (Y_1)	0.65* (0.15)	0.21 (0.18)	—	0.86++	0.67
Andhra (1958-59)	16	Milk (Y_1)	0.69** (0.24)	0.45+ (0.30)	—	1.14++	0.68
Andhra (1959-60)	15	Milk (Y_1)	0.89* (0.21)	0.22 (0.42)	—	1.11++	0.65
Punjab (1954-55 to 1956-57)	15	Dung (Y_2)	—	—	0.41** (0.15)	0.41	0.30
Andhra (1957-58)	15	Dung (Y_2)	0.74* (0.26)	—	—	0.74	0.38
Andhra (1958-59)	16	Dung (Y_2)	0.20** (0.27)	—	—	0.20	0.37
Andhra (1959-60)	15	Dung (Y_2)	0.01 (0.02)	—	—	0.01	0.07

* Significant at 1 per cent. ** Significant at 5 per cent.
\+ Significant at 10 per cent. ++ Not significantly different from unity.

Note: Y_1 = Milk yield per milch animal (in *maunds* for Punjab and in *seers* for Andhra) per annum.
Y_2 = Value of dung yield (Rs.) per milch animal per annum.
X_1 = Value of feed (Rs.) per milch animal per annum.
X_2 = Value of labour input (Rs.) per milch animal per annum.
X_3 = Value of all inputs (Rs.) per milch animal per annum.

Notes

1. V.M. Dandekar, (1969), "India's Sacred Cattle and Cultural Ecology," *Economic and Political Weekly*, September 27.
2. *Ibid.*
3. *Ibid.*
4. K.N. Raj, (1969), *Investment in Livestock in Agrarian Economies: An Analysis of Some Issues Concerning 'Sacred Cows' and 'Surplus Cattle'*, Delhi School of Economics.
5. Dandekar, *op.cit.*
6. S.N. Mishra, (1966), "Cattle-Meat and Economic Welfare," *Kyklos*, Vol. XIX, Fasc I.
7. S.C. Chaudhri and R. Giri, (1963), "Role of Cattle in India's Economy," *Agricultural Situation in India*, December, p. 596.
8. V.M. Dandekar, (1964), "Problem of Numbers in Cattle Development," *The Economic Weekly*, Annual Number, February, p. 353.
9. ———, (1969), "Cow Dung Models," *Economic and Political Weekly*, August 2.
10. K.N. Raj, *op. cit.*
11. Government of India, *Report on India's Food Crisis and Steps to Meet it* by the Agricultural Production Team Sponsored by the Ford Foundation Team, 1959, p. 64.
12. Directorate of Economics and Statistics, Ministry of Food and Agriculture, *Studies in the Economics of Farm Management*, "Punjab (1954-55, 1955-56 and 1956-57) and Andhra Pradesh (1957-58, 1958-59 and 1959-60)."

Economic and Political Weekly, Vol IV, No. 52, Review of Agriculture, December 1969.

14

India's 'Surplus' Cattle: Reply

In his article "Sacred Cattle and More Sacred Production Functions"[1] V.M. Dandekar raises two important questions. First, he observes that for a test as to whether there exists surplus cattle in relation to the given resources, what is relevant are not output elasticities for an animal with respect to resource inputs as formulated by me[2] but marginal productivities of inputs. Secondly, he argues that the choice of the Cobb-Douglas production function is inappropriate for this purpose.

Let me start with the first point which, I think, is more important than the second one. It would be useful at this stage to spell out the issue under discussion before I proceed to examine the relevance of output elasticities in resolving the issue. K.N. Raj asked the question "whether, given the amount of feed that can be spared for maintaining animals in the situation, a higher output is secured by giving this food to a smaller number and raising their productivity or by distributing it over a larger number" and suggested that "a firm answer to this question cannot be offered without much deeper investigation into the relevant production functions."[3] Dandekar raised the same issue as is evident from the following: "with better feeding their productivity can be more than proportionately increased; that, therefore, with given resources, a smaller stock of such animals can give more milk, traction and dung than the much larger stock of non-descript animals at present gives."[4] That this is exactly the issue under discussion is reiterated by Dandekar in his criticism[1] of my article by quoting the above passages from Raj and himself.

I thought that the issue raised by Raj and Dandekar is a relevant one but since there is hardly any empirical evidence bearing on the range of the production function at which the cattle are actually maintained in rural India, I proceeded to make an attempt in this direction by translating the Raj-Dandekar formulation into the following testable proposition: "If a significant female cattle population is surplus or redundant then one should expect the elasticity of output (*i.e.*, breeding efficiency, milk and dung) response for an animal with respect to the feed and other inputs to be more than unity. Elasticities significantly less than unity could mean that the existing number is less than the number desired by the farmers on economic grounds."[2] Dandekar says that "this is not a correct formulation; for what is relevant here are not elasticities but marginal productivities."[1]

Dandekar does not demonstrate how output elasticities are irrelevant or how marginal productivities are in fact relevant for answering the question raised by Raj and himself. If Dandekar wanted to disprove my proposition he should have demonstrated how output from the *given* resources is not maximised when the output elasticity for an animal is unity. He should have specified his marginal conditions corresponding to maximum output from the *given* resources, and should have demonstrated how the output elasticities specified by me fail to correspond with his marginal conditions when output is maximum.

I shall now try to demonstrate with the help of the Figure 14.1 how output elasticities are relevant to answering the question under discussion. OX axis in the Figure represents input per animal in rupees and OY axis measures the output per animal in rupees. The total product curve (TP) for an animal represents an initial range of increasing returns where the output elasticity is more than unity (e>1), then the point where the output elasticity is unity (e=1) followed by a range of decreasing returns with output elasticity less than unity (e<1) before the elasticity falls off to zero. The slope of the line OC is unity, *i.e.*, for a rupee increase in input per animal, output per animal increases by a rupee. Line O_1C_1 being parallel to OC has the same slope. At y_2 the slope of the TP curve is equal to the slope of O_1C_1, that is, marginal productivity is unity at this point. The line OC_2 is tangent to the TP curve at y_1, so that when the TP curve reaches this point, average product curve (AP) would reach its maximum point (m) at which it would

Figure 14.1

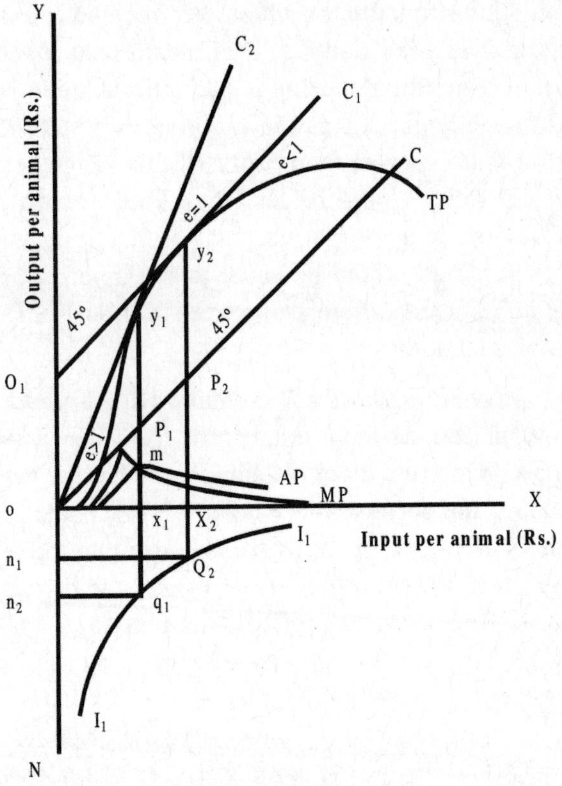

intersect with the marginal product curve (MP). Thus output elasticity would be equal to unity (e=1) at point y_1. The axis ON to the south represents the number of animals that can be maintained with the given total input at varying input levels per animal. Curve I I_1 is the equal input curve relating input per animal with the number of animals that can be maintained with the given total input. For instance, when input per animal is OX_2, the number of animals would be On_1. When the input per animal is OX_1, the number of animals that can be maintained with the same total input would rise to On_2. The area of the rectangle $Ox_2q_2n_1$ is thus equal to that of $Ox_1q_1n_1$.

Dandekar says that "an animal should be maintained, if it is maintained for an economic purpose, within the middle range of the production function in which the marginal productivity of feed is greater than unity".[1] This means that Dandekar is in favour of extending the input per animal even beyond X_1, though not beyond X_2, as the marginal productivity of input would still be greater than unity in this range, even though output elasticity would be less than unity. Profit per animal would rise so long as the marginal productivity is more than unity. Since X_2P_2 would be equal to the input OX_2, profit per animal at X_2 or the difference between X_2y_2 and X_2P_2 (*i.e.*, y_2P_2) would be maximum.

But maximum profit per animal will not yield maximum total output or maximum total profit from the *available* input resources. Given the input resources, whether for the individual farm or for the society as a whole, the input rate (per animal) that maximises total output will maximise total profit also. The maximisation of output *per rupee of input* will maximise total output as well as total profit.[5] This point (*i.e.*, the point of maximum average productivity of input) is reached when the output elasticity is unity. This is the point where the average product curve intersects with the marginal product curve. As I have suggested, below the input level (per animal) where output elasticity is unity, the actual number of animals would be more than the desired number. Therefore, output can be increased by reducing the number of animals (increasing the input per animal) upto the point where output elasticity equals unity. Beyond that, although marginal productivity would still be greater than unity (upto the point y_2 on the TP curve), the elasticity is less than unity and the average product would be declining so that total output from the given inputs would be declining indicating that the number of animals is less than the desired number on economic grounds. It is clear that Dandekar's position would result in a smaller number of animals (larger input per animal) than that implied in my argument. By allocating the available resources to a number that is less than optimum, Dandekar's prescription would fail to maximise output and profit.

It is thus evident that Dandekar's marginal productivity criterion is not consistent with his earlier proposition, *viz.*, output from the available resources can be increased by reducing the number of animals and increasing the input per animal. In fact, my formulation that output elasticity needs to be more than unity for increasing output from the given resources is consistent with his earlier proposition.

On the second point, *viz.*, the choice of the Cobb-Douglas production function, Dandekar raises three objections. Firstly, he says that "the elasticity for any one factor such as feed cannot be anywhere near unity and that the larger the number of factors included in the production function, the smaller will be each single co-efficient."[1] But I did not suggest that we should expect each single co-efficient to be more than unity if surplus cattle exists. My formulation, as quoted by Dandekar himself, says that "one should expect the elasticity of output response for an animal with respect to feed and other inputs to be more than unity."[2] Thus I am looking for the sum of elasticities and not the individual elasticities. This is evident from my conclusion that "in Andhra the actual number seems to have been close to the desired or optimal number"[2] because the sum of elasticities is not significantly different from unity, even though the individual elasticities are less than unity. The reason why I am looking for the sum of elasticities is that apart from fodder and feed, labour is an important input for cattle. As indicated in my article (Table-13.1, page 178) small farmers who have limited feed resources seem to be using more labour for the maintenance of cattle when compared to large farmers whose feed resources per animal could be higher and who have to hire labour. Since labour input cannot be considered costless especially for the large farmers, it would only be proper for an empirical test such as this to take account of feed as well as labour inputs when the sample observations include small as well as large farms.

Dandekar's second objection seems to be that since production functions presented by me are per animal "the sum of the elasticities in the Cobb-Douglas production function would not be close to unity", so that the scale co-efficient will not provide an efficient test. This can be dealt with along with his third objection to the use of the Cobb-Douglas production function, *viz.*, "it assumes the elasticities to be constant". It is true that when sample observations or experimental data are clustered around the different ranges of production elasticity, the use of the Cobb-Douglas production function is not useful because, owing to its constant elasticity property, it cannot yield the full story but only an average position. However, the scatter of observations we have used does not indicate varying ranges of elasticity. Besides, we are concerned here with a situation where it is contended that bulk of the cattle are under-fed at present and that, therefore, a smaller number of better-fed animals will contribute more output than at present. If this is true, one should expect bulk of the observations to lie around the range where the output elasticity is more than unity and there is no *a priori* reason why the Cobb-Douglas production function cannot reflect this situation and yield the sum of elasticities which is more than unity, even though the elasticity has to be constant within this range.

The above is meant only to suggest that the Cobb-Douglas production function cannot be dismissed entirely for dealing with the issue under discussion. However, there is no sanctity about this particular type of the function and I am one with Dandekar on the need to experiment with other and more appropriate types of functions. This would require better data and much larger number of observations than I was able to use. In fact, I had suggested at the end of my article that "more intensive investigations and with better data for different regions of the country are needed before firm conclusions can be drawn."[2]

Dandekar strikes the following pessimistic note so far as empirical work is concerned: "Judging by the difficulties of securing data and deriving necessary production functions therefrom listed by Hanumantha Rao, it seems that we shall never have a satisfactory set of production functions in our armoury adequate to answer the question."[1] However, despite these difficulties Dandekar feels that "... we know enough about the nature of the relevant production functions and the state of the cattle in the country to be able to answer certain questions of practical policy in this field. As mentioned above, we can reasonably guess that the production function will have an initial range where the marginal productivity rises from zero to unity, a middle range where the marginal productivity rises above unity and falls back to unity, and an end range where the marginal productivity declines from unity to zero".

But is this *a priori* knowledge on the nature of the production functions sufficient to warrant policy prescriptions on a vital question like this? I had suggested that the question under discussion "can be settled only by investigating the nature of the production functions underlying the Indian cattle economy".[2] Dandekar makes a similar suggestion: "we must know not only the production functions appropriate to different breeds and feeding practices but also as to how many animals are maintained at different points of the production function in each case."[1] Thus what we need to know for this purpose is the evidence as to the *points* or the *ranges* on the production function at which Indian cattle are *actually* maintained. No policy prescriptions regarding the size of the cattle would seem permissible in the absence of such evidence.

Appendix*

Assume that cows are divisible. Let f(x) be the output per cow when it is fed x. Let the total number of cows held by the peasant be $\overline{N}$ and total feed held by him be $\overline{X}$. Assume further that the farmer cannot buy or sell feed nor can he buy cows. However, he can reduce the head to any size he likes costlessly. Then the two decision variables of the farmer are: N the size of the herd and x_i the amount of feed to be given to the *ith* cow. Assuming that all cows are alike, it is clear that the same amount of feed will be given to each cow. Hence the decision variables reduce to N and x, the feed per cow. The constraints are $0 \leq N \leq \overline{N}$ and $0 \leq Nx \leq \overline{X}$. Let f(x) be the output of a cow when it is fed x. The farmer wishes to maximise Nf(x), *i.e.*, the total output of his herd, subject to $0 \leq N \leq \overline{N}$ and $0 \leq Nx \leq \overline{X}$. In conformity with the received tradition let us assume that: $f(0)=0$, $f'(x)>0$ for $0<X<X^*$ and $f'(X^*)=0$. Let $f''(X)>0$ for $0<X<X^{**}$ and $f''(X)<0$ for $x^{**} < X \leq X^*$. Let $f''(X^*)=0$. Thus the relevant range of feed is $[0, X^*]$. Output per cow is maximised at X^*. Marginal product of feed is maximised at X^{**}. There is another value $\overline{x}$, where average product per unit of feed, *i.e.*, $\frac{f(X)}{X}$ is maximised. It is clear that at

* The author is indebted to T.N. Srinivasan for this note.

$\tilde{x}, \frac{d}{dX}\left[\frac{f(x)}{x}\right]=0$, *i.e.*, $\frac{\tilde{x}f'(\tilde{X})-f(\tilde{X})}{\tilde{X}^2}=0$ or $\frac{\tilde{X}f'(\tilde{X})}{f(\tilde{X})}=1$, *i.e.*, elasticity of output per cow with respect to feed is unity and $\frac{d^2}{dX^2}\left[\frac{f(X)}{X}\right]<0$, i.e., $\frac{f''(\tilde{X})}{\tilde{X}}<0$. This second order condition implies $\tilde{X}>X^{**}$. This first order condition implies $f'(\tilde{X})>0$ and hence $(\tilde{X})>X^*$. Thus $X^{**}<\tilde{X}<X^*$. Assume that $\frac{f(X)}{X}$ has a unique maximum at $\tilde{X}$.

Let us turn now to the farmer's maximisation problem. It is obvious that it will never pay the farmer to feed each animal anything less than $\tilde{X}$. To prove this, suppose the farmer has chosen the herd size $N(\le \overline{N})$ and a feed level $X=\frac{\overline{X}}{N}<\tilde{X}$. Consider the alternative decision, $N'<N\le\overline{N}$ and $\tilde{X}$ where $N'\tilde{X}=\overline{X}$.

Now $\frac{f(\tilde{X})}{\tilde{X}}>\frac{f(X)}{X}$, *i.e.*, $\frac{f(X)}{\frac{\overline{X}}{N'}}>\frac{f(X)}{\frac{\overline{X}}{N}}$ *i.e.*, $N'f(\tilde{X})>Nf(X)$,

i.e., farmer gets a larger output by reducing the size of the herd to N' and increasing the feed to $\tilde{X}$. Can it hapen that the farmer feeds more than $\tilde{X}$ to his herd? This can happen if he has enough feed $\tilde{X}$ so that even distributing it over his entire herd, he gets an average feed $\frac{\overline{X}}{\overline{N}}>\tilde{X}$. Of course the farmer will never choose to feed more than X^* per cow. Hence we can describe the farmer's decision as follows:

$$A: \frac{\bar{X}}{\bar{N}} < \tilde{X}; \text{decision } X = \tilde{X};$$

$$N = \frac{\bar{X}}{\tilde{X}} < \bar{N}.$$

Hence available feed is fully utilised and optimal herd size is below the available number $\bar{N}$. Output Q in this case is

$$\frac{\bar{X}}{\tilde{X}} = f(\tilde{X})$$

$$B: \frac{\bar{X}}{\bar{N}} = \tilde{X}; \text{decision } x = \tilde{x} \text{ and}$$

$$N = \bar{N}; Q = \bar{N}f(\tilde{x}) = \frac{\bar{X}}{\tilde{x}} f(\tilde{x})$$

$$C: X^{*} \geq \frac{\bar{X}}{\bar{N}} > \tilde{x}; \text{ decision } x = \frac{\bar{X}}{\bar{N}}$$

$$\text{and } N = \bar{N}; Q = \bar{N}f\left[\frac{\bar{X}}{\bar{N}}\right]$$

$$D: \frac{\bar{X}}{\bar{N}} > X^{*}; \text{ decision } x = x^{*} \text{ and}$$

$$N = \bar{N}; Q = \bar{N}f(x^{*})$$

Note

* The author is greatful to T.N. Srinivasan for the helpful comments and suggestions on an earlier version of this note. However, responsibility for the views expressed and for errors is entirely his.

References

1. V.M. Dandekar (1970). "Sacred Cattle and More Sacred Production Functions," *Economic and Political Weekly*, March 21.
2. C.H. Hanumantha Rao (1969). "India's 'Surplus' Cattle: Some Empirical Results," *Economic and Political Weekly*, December 27.
3. K.N. Raj (1969). *Investment in Livestock in Agrarian Economics: An Analysis of Some Issues Concerning 'Sacred Cows' and 'Surplus Cattle',* Delhi School of Economics.
4. V.M. Dandekar (1969). "India's Sacred Cattle and Cultural Ecology," *Economic and Political Weekly*, September 27.
5. Earl O. Heady, *Economics of Agricultural Production and Resource Use*, Prentice-Hall, 1952; See especially pages 77, 78, 116, 118, 127 and 128.

Economic and Political Weekly, Vol. V, No. 40, October 3, 1970.

15

Agriculture in India and China

A Comparison

The Period of Sixties and Seventies

Despite the basic differences in social systems, the growth of agriculture was strikingly similar in India and China, although for very different reasons. Throughout the sixties and seventies, foodgrains output grew at the rate of around 2.5 per cent per annum both in India and China; and because of population growth of a little over 2 per cent per annum, improvement in per capita consumption was negligible in both the countries. However, the initial level of foodgrains output — per hectare as well as per head — in China was substantially higher than in India. Thanks to radical land reforms and other institutional changes, the inequality in consumption has also been much lower in China.

The impact of weather on foodgrains output has not been as adverse in China as in India, because, nearly half the cultivated area is under irrigation in China as against only about 30 per cent in India. However, thanks to the democratic political framework within which decision-making is done and the operation of markets and incentives to the producers, there has been a steady relationship between the growth of agriculture and the rest of the economy in India. In China, on the other hand, policy-induced instability was considerable, causing dislocations in the economy and hardships to the people. During the period of Great Leap Forward, for example, agricultural output declined at the rate of nearly 4 per cent per annum continuously for four to five years resulting in severe famine in the early sixties.

Strangely enough, this happened not because of neglect of agriculture but because of over-attention to it: over-zealous targets were set and input-use was intensified without any regard for the law of diminishing returns which was condemned as a bourgeois law! The plant population and fertiliser-use per hectare were stepped up enormously through the arbitrary choice of seed varieties and cultural practices. Because of such ideological interference with Science and Technology and disregard for decentralised decision-making by farmers, the productivity of resources in agriculture declined significantly. In contrast, inefficiencies in resource-use in Indian agriculture prevailed not so much at the farm level as at other levels where bureaucratic management is involved e.g., in the execution of irrigation projects, delivery of inputs and the provision of extension services etc.

India seems to have allocated a larger proportion of its public outlays for agriculture than China. However, the overall rate of capital formation in agriculture has probably been higher in China. This is because the rate of accumulation in Chinese economy has been much higher, i.e. around one-third of gross national product, as against around 25 per cent reached in India in the recent period. The prevailing political framework in China enabled it to mobilise labour in a big way for capital construction, e.g., small scale irrigation projects, soil conservation, roads and other infrastructure, apart from the extensive use of labour in the preparation of organic manure which accounted for nearly half of the plant nutrients used in China.

The terms of trade between agriculture and the rest of the economy have, by and large, remained steady in India throughout the plan period except that there has been some decline in relative prices recently in the case of crops like wheat and rice which have experienced technological changes and rise in productivity. In China, terms of trade have been deliberately shifted against agriculture until recently with a view to transferring resources to the non-agricultural sector. Similarly, direct taxation of agriculture has been sizeable in China, whereas it has been negligible in India. Thus, the system of incentives for farmers has been weaker in China. And, owing to the assured and more or less equal wage regardless of the nature of work, productivity of labour in terms of effort put in had gone down considerably.

These dislocations and inefficiencies notwithstanding, the Chinese system has been able to ensure much greater equity in the distribution of benefits from agricultural growth than in India. This is because of social ownership of means of production including land, guaranteed employment and fair wage. The proportion of rural population below the poverty line was much lower than in India, despite the restrictions on rural-urban migration. Whatever poverty that exists, is concentrated in the backward regions, even though the regional disparities in development are less than in India. The regional disparities have been held down partly by policies for regional self-sufficiency which required that each region produced certain minimum quantities of grain and other commodities regardless of resource endowments. Also, adverse terms of trade for agriculture and restriction on markets prevented the well-endowed regions from moving faster. Among positive factors, the successful implementation of land reforms and the massive mobilisation of labour for capital construction throughout the country-side and the diversion of part of the surpluses of the developed regions to the neighbouring less developed regions have contributed to reducing disparities in regional development.

Internal migration in India, both rural-to-urban and rural-to-rural, has undoubtedly worked to lessen the rural poverty. But poverty in India is only partly attributable to uneven levels of regional development. It arises essentially from the unequal access to resources including land, unemployment and low wages and various forms of exploitation of the poor.

Learning from the experience of two decades, the Chinese have brought about major shifts in agricultural policy since 1978. Agricultural prices have been raised significantly relative to those of industrial goods; there has been massive injection of capital into agriculture; inter-regional trade has been opened up; and communes have given place to responsibility system based essentially on household enterprises where earnings are related to performance. The opening up of markets and operation of incentives to farmers resulted in a big spurt in agricultural growth in the last few years. Improved use of hitherto underutilised infrastructure, regional specialisation in production and inter-regional trade and enhanced labour productivity are the major sources of agricultural breakthrough experienced recently. This process is no doubt giving rise to some disparities between different regions and households but the major impact has been to release productive forces, use resources efficiently and improve the standards of living.

Interestingly, both in India and China, further agricultural growth is limited by demand, although the factors affecting demand are not the same in the two countries. With high accumulation rates and low disposable income and population growth coming down drastically in the recent period, demand has become critical for further growth of foodgrains output in China. Already, accumulation rates are being brought down and there is a move towards greater diversification of agriculture to cater to rising incomes.

The prospects thus appear bright, but the Chinese political system, despite all its virtues, makes for highly doctrinarian and centralised decision-making with all its pitfalls for agriculture.

Policy-instability is inherent in the system which also ensures thorough-going implementation of policies which are later proved to be manifestly wrong.

In contrast, the Indian political system provides adequate safeguards against the sudden emergence of wrong policies and certainly against their effective implementation. The demand problem in India arises from the inadequate generation of purchasing power for the poor. This needs to be tackled through an appropriate mix of employment-oriented production strategy and poverty alleviation programmes.

The Eighties: Shift in Policies in China*

There has been intensive drive towards modernisation of economy and increasing the productivity of labour and capital through modernisation in all areas whether it is industry or agriculture. This is being achieved through decentralisation of their economic decision-making. Much more power is being now given to the units concerned. Decisions about investment and production are being made at lower levels. It is at these levels that they decide how much labour to engage and how much to pay for them.

Secondly, they are trying to improve the motivation of labour. For a long time they have had the ideal of an egalitarian system, by which they meant more or less equal payment for all kinds of labourers without much regard for the quantity and quality of work actually being put in by them. Now, they are relating wages to the actual work done, that is to say, they are building incentives into the system of wages. This way they want to improve the productivity of labour. These are the two major ways by which they want to improve the productivity.

---

* Author's impressions of the changed economic scenerio when he visited China as leader of the Indian delegation sent by ICSSR in Oct., 1986 to participate in the joint ICSSR-CASS Seminar on Planning. The text is based on his interview to *Patriot,* November 16, 1986.

One can not say whether this marks a basic shift in the ethos of the society. But, there is certainly a shift from their practice during the last two decades or so especially between 1957-77. They were then engaged with institutional restructuring. Perhaps, they were trying to achieve the communist goal of ensuring wages for everybody according to needs. To use the Marxist terminology, perhaps they did not really wait for the productive forces of technology to mature and went ahead rather fast with the restructuring of production relations. Now, perhaps, they realised that they had been going too fast with the restructuring of production relations without much regard for improving the productive forces, the technology, and, therefore, now they are concentrating more on developing the productive forces and wanting to proceed gradually on changing the production relations.

Thus there seems to be a slight correction in their priorities. However, the Chinese system is still very egalitarian as far as the distribution of income is concerned although they don't want to use the term now. From our standards it is far more egalitarian. There is no private ownership of land. All the other means of production in the non-agricultural sector are also owned by the state. Basically the operation of the means of production, the sharing of the product and the public services, are still regulated by the state. They are very different from the usual market economies.

After their two decades of experience they have felt the need to correct the system with a new emphasis on productive forces.

The industries are totally in the public sector. They realised that at least half of them had outmoded equipment which needed modernisation; wherever necessary they are importing technology. But it is highly selective in nature. The thrust is primarily on the indigenisation of technology they are importing.

They are also conscious of its effects on employment and they have repeatedly said that their enterprises have to remain labour-intensive for some more time to come. The unemployment rate is not more than 1 per cent in urban as well as rural areas and, according to them, there is no unemployment. Yet they are still very conscious of the implications of technological change as far as employment potential is concerned.

Much of the technology that is coming is from Japan although they are borrowing from all the countries of the world. They even said in one of our meetings that they would like to have increased collaborations with India. They openly acknowledged that India is ahead of them in quite a few sectors in the technological field.

Self-reliance remains the major objective. Despite all these collaborations, the foreign resources component as a proportion to the total investment in the economy is not more than two per cent now. As a part of the decentralisation policy, the enterprises now have the freedom to import technology on their own without prior clearance from the central authority. They admit this will lead to an excessive drain of foreign exchange in some cases. But they are confident that they can correct it; they have the necessary instruments to restore the priorities.

They have joint ventures also with foreign companies of the private sector there. In fact, they have a number of special economic zones, like the one we visited, Shenzhen near Hong Kong, the South of China. There are all kinds of joint ventures with many countries of the world. This is designed to augment their foreign exchange partly, but mainly this is designed to import technology, create employment opportunities and improve the skills of their local labour.

For some time they allow repatriation of profits. But the controls will be put if felt necessary. Despite the outflow of foreign exchange for inputs there is a net earning of foreign exchange for the country. Though it is not very significant, it is positive.

China is also embarking upon a radically new labour policy. First of all, they want to develop a labour market which did not exist earlier. Everybody had life-time employment and employment was regulated by the state. Now the enterprises will be free to hire the kind of labour they want. They will also be free to fire.

But there are also certain safeguards attached to it. They announced a new labour policy to be implemented from 1st October 1986 when we were there. The employment has now become contractual. All new employment will be contractual for five years and labour can be thrown out after that. But they have to guarantee certain minimum wages until they are re-employed and there is also a provision for retraining of labour.

They are proceeding cautiously. They first want to make substantial improvement on the existing position. For a long time many of the public enterprises in China were allowed to operate with a low technology which has become obsolescent. They realise now that some of these enterprises need to be modernised. But there is also the realisation that the labour-intensive nature of industries cannot be done away with.

Modernisation in Agriculture

The reform in agriculture is far reaching. For about 20 years or so in the communisation period, the exchange of agricultural commodities between different regions was almost nil. The communes were virtually self-sufficient in every respect. Now, they are being opened up. Inter-regional trade is being allowed. This has resulted in each region beginning to specialise in certain types of crops.

They are also now giving less emphasis to cereal foodgrains and more to non-cereal food like dairy, poultry etc. They are diversifying their agricultural economy and are inducting new technology and beginning to concentrate on agro-based industries. And the trend of decentralisation can be seen much more visibly in agriculture. They are reverting from the communes to the household as the unit.

The decision is related to the productivity of labour. For nearly two decades the labour productivity was stagnating because everybody was getting the minimum and, as they say, they were eating from the common pot. There was no incentive to work hard because if you worked hard you did not have additional remuneration. May be with the first generation of revolutionaries immediately after the revolution in the fifties, there was no such problem. But later this problem of incentives arose and then they had to revert to the system where the individuals have an incentive to work hard.

In India, egalitarianism is a very respectable word although we may not be practising it. But in China, surprisingly, it has become a dirty word. In fact they admit that they went wrong by practising egalitarianism and want to get away from it.

Now, they admit that this kind of decentralisation in rural areas and emphasis on household and on market mechanism will lead to increase in disparities. In fact it has already led to that. Households doing well are infact being given bigger plots to work on.

At the moment their slogan seems to be — let some people get rich so that others can follow. It is unprecedented because we never find this in communist literature in any other communist country; with all their variants of revisionism, we never find such a slogan.

Perhaps in China they were so sick of the so-called egalitarianism and the resultant low labour productivity, that they decided to embark on this path. They admit that the disparities are increasing but according to them, these can be redressed in course of time.

There has been mechanisation of various kinds for quite some time in Chinese agriculture. And it is highly regulated. It is not the kind which is leading to unemployment. In fact, mechanisation has been necessitated in the Chinese situation by the scarcity of labour. They have allocated some labour in agriculture and moved some labour already to non-agricultural enterprises. So wherever there were shortages, mechanisation had to be undertaken.

Intellectuals seem to be the happiest lot. Most of them were very sore with the period of cultural revolution. Their position was relegated to background. They think that they were humiliated. Most of them were sent to rural areas to do manual work. We met a professor who had been made to sweep the University floors under the supervision of his students. Now the authorities realise that a whole generation in that period had been rendered less useful. As they did not have proper training, they did not learn anything. That is seen as the cause of declining productivity.

16

Reforming Agriculture in the New Context

The Institutional Context

The organisation of Indian agriculture has been based on individual farming under the system of peasant proprietorship, unlike in the former socialist countries. Moreover, no attempts have been made in India to deliberately shift the terms of trade against agriculture. On the other hand, owing to the persistence of imbalances between the supply and demand for agricultural commodities, particularly foodgrains, the terms of trade have continued to be favourable to agriculture throughout the greater part of the plan period including the first decade of green revolution (Thamarajakshi 1977). Also, because of growing influence of farmers, inevitable in a democratic polity, the farm subsidies — open as well as hidden — have been mounting over a period of time undermining the capacity of the government to make productive investments in agriculture. The rise in subsidies on irrigation and fertilisers has induced the allocation of resources in favour of crops like wheat and rice resulting in the achievement of self-sufficiency in foodgrains and, consequently, some decline in the relative price of these crops during the 80s.

There was considerable under-utilisation of capacity in agriculture in the former socialist economies both because of very high rate of capital accumulation for over four decades, achieved in part by using surplus labour for capital construction as in China, and because of the organisation of agriculture under the system of collectives and communes leading to the suppression of markets and gross inefficiencies in the use of resources. In China, for instance, in total disregard for agro-climatic variations across regions, the policy of self-sufficiency in agricultural commodities in each region was strictly enforced. The authorities in communes used to decide what seed to be sown and how much fertilisers to be applied. Therefore, there was a big spurt in agricultural output in China in the eighties when the communes were disbanded, inter-regional trade was opened up and terms of trade were shifted in favour of agriculture.

In India, on the other hand, the rate of capital formation in agriculture has been much lower and utilisation of available capacities relatively more efficient. For example, the proportion of cultivated area irrigated in India is only about one-third, as against over 50 per cent in China. The level of human capital formation in terms of literacy, education and health of rural population was also much higher in China when economic reforms were launched. Moreover, except for occasional restrictions on the inter-regional movement of foodgrains for the purposes of public procurement in India, there has, by and large, been a free flow of commodities and resources between different regions in the country. The success of green revolution would have been unthinkable in the absence of integrated national market in India which has been conducive to regional specialisation in the production of crops. Therefore, unlike in the socialist or command economies, there is little scope for stepping-up agricultural output in India merely by improving the efficiency of resource-use at the farm level by restructuring the management of agriculture or by redefining the role of the state in relation to markets.

Reforms in the Indian Context

The economic reforms for Indian agriculture at present need to be directed basically towards augmenting the productive potential rather than designed merely to improve the utilisation of available resources. The declining trend in investments in agriculture, increasing land degradation, technological stagnation and growing non-viability of credit institutions reinforce the need for such reforms. The reforms for agriculture that need to be undertaken fall broadly under two categories, *viz.*, infrastructural-technological and institutional. The three major areas of concern under the former category are: (i) stepping up public investment in agriculture, among other things, by reducing farm subsidies; (ii) exploring new avenues of technological breakthrough such as the introduction of biotechnologies; and (iii) strengthening the programmes for soil and moisture conservation in dryland areas by integrating wage-employment programmes with them. The areas of concern on the institutional front are: (i) restoring the health of rural credit institutions; (ii) restructuring institutions for decentralised management of agricultural infrastructure; and (iii) raising incentives for investment and effort by the farmers and improving the efficiency of resource-use by relaxing restrictions on trade in agricultural commodities.

Farm Subsidies Versus Investment

It would be instructive to know the increase in the magnitude of subsidies for agriculture incurred by the centre and states together in relation to plan expenditure for the development of this sector. For irrigation we are considering only the subsidy incurred on account of the non-recovery of operation and maintenance charges leaving out the cost of capital deployed. Further, based on the world price of fertilisers, we are assuming that only about half the actual subsidy on fertilisers benefits the farmers and the consumers (Gulati 1990).

The input subsidies, so calculated, on fertilisers, irrigation, electricity and credit combined constituted around one-third of total plan expenditure on agriculture, irrigation (including flood control) and Special Area Programmes incurred by the centre as well as state governments in 1980-81, and rose steeply to as high as 90 per cent of such plan expenditure in 1989-90 (Rao 1994). If annualised cost of capital invested in irrigation is also accounted for at historical cost, which is quite modest when compared to the replacement cost, the subsidy on surface irrigation during the Seventh Plan period (Rs. 11,800 crore) exceeded the Seventh Plan expenditure (Rs. 11,000 crore) on major and medium irrigation. The subsidy on rural electricity during Seventh Plan period amounted to Rs. 11,700 crore which is about three times the Eighth Plan outlay of Rs. 4,000 crore provided in the State Plans for rural electrification in the Power Sector (Eighth Five-Year Plan, p. 197). It is clear, therefore, that for every rupee of investment in agriculture, we are incurring today more than a rupee by way of subsidies on inputs for agriculture.

These subsidies have reached a point where their beneficial effects seem to be outweighed by the adverse effects in terms of macroeconomic imbalances, slowing down of public investment in agriculture, inefficient use of resources at the farm level, degradation of environment, and reduction of employment. Since there is now an over-use of these inputs in limited pockets, efficient pricing of such inputs may lead to greater diversification of their use benefiting a larger number of crops over wider areas. This would result in greater employment and output per unit of capital invested.

The latest data from the CSO up to 1990-91 shows that although gross capital formation in agriculture by the private sector at 1980-81 prices has shown an increase since 1987-88, that by the public sector continued to decline significantly, so that the total real gross capital formation in agriculture in the late-eighties is still lower than that achieved in the early-eighties. This trend has to be reversed by stepping up public investment in agriculture. For this to happen, it is imperative to reduce subsidies on inputs. However, there is considerable political resistance and, in a democratic polity, the process of change towards rational pricing of inputs is bound to be slow. In these days of globalisation and easy flow of information, it is becoming difficult to convince our farmers about the need to reduce such subsidies when their counterparts in the market economies like the US, Japan and European Community enjoy much higher level of subsidies. Despite some reduction in fertiliser subsidy, the plan investment in agriculture has been declining because of the negligible rise in central assistance to state plans on account of the drive for reducing fiscal deficits. This is being attempted not by reducing revenue deficits but largely by reducing capital expenditures. Also, the states' resources have been eroded by the slow rise in the revenues from the shareable taxes like income tax and union excise duties consequent to the recent reduction in tax rates.

Technological Upgradation

The introduction of biotechnology may contribute to the protection and regeneration of environment by reducing the dependence on chemical inputs. If farm subsidies are reduced significantly in the developed countries in pursuance of the recently concluded GATT agreement, chemical fertilisers may increasingly be replaced by the renewable sources like biologically fixed nitrogen (Jain 1993).

However, being knowledge-intensive and skill-intensive, biotechnologies necessitate greater investment in research as well as in improving the capabilities of farmers. Although some advanced centres of research in biotechnology have been established in the country, it is not clear how such priorities find their place in the existing agricultural research system in India.

It is necessary to have a fresh look at the priorities of the Indian agricultural research system in the light of these emerging prospects and for adequately taking into account the socio-economic factors relevant in the evolution of new technologies.

India is currently investing only about 0.3 per cent of its agricultural domestic gross product in agricultural research, as against 0.7 per cent in the developing countries as a whole and as much as 2-3 per cent in the developed countries (Jain 1993). Therefore, so far as India is concerned, there is considerable scope for diverting incremental outlays for the priority areas in research. Strengthening indigenous capabilities in biotechnology by involving the private sector in research as well as in the multiplication and distribution of seeds provides an effective safeguard, in the long-run, against the patenting of seeds and other restrictive practices by the multi-national corporations.

Soil and Moisture Conservation

A large part of unirrigated area is characterised by low and uncertain rainfall, low wages and high poverty. Owing to the unregulated exploitation, these fragile areas have been highly degraded. Because of low and defective pricing of electricity, the pumping of ground water has been greater than the recharge resulting in low water table and shortage of drinking water. Nevertheless, there is considerable scope for soil and moisture conservation and for harvesting rain water in such areas by arresting the run-off. The ongoing programmes for this purpose need to be strengthened through better planning and sequencing of works on a watershed basis by involving people and improving co-ordination between various departments. The wage employment programmes, so essential for these areas, can be integrated with the ongoing programmes for soil and moisture conservation, as is being attempted through the Employment Assurance Scheme (EAS) recently launched by the government to guarantee wage employment for 100 days for those in need and seeking employment in drought-prone, desert, tribal and hill areas.

Some advocates of economic reform feel that such employment programmes are populist, as they may result in raising the fiscal deficit (Narasimham 1993). It is important to recognise that such programmes, when properly implemented, will raise agricultural productivity, improve equity and promote sustainability. Indeed, in a democratic polity like ours, reconciling the goals of efficiency with those of social justice is necessary for sustaining the reform process itself.

Credit Reform

According to the Report of the Agricultural Credit Review Committee (RBI 1989), the rural credit system is now threatened by rising overdues and deficiencies in management attributable basically to the policy environment as well as political culture which encourage lending at subsidised rates for activities that are not productive, discourage professional approach and necessary follow-up in lending, and promote wilful default on the part of those who have the capacity to repay. As a consequence, recycling of credit has become a major casualty, the losers being the present as well as the prospective borrowers. Therefore, improving the viability of credit institutions should be foremost on the policy agenda.

Institutional credit for agriculture has to expand at a faster rate than before because of the need to step up agricultural exports and also because the changing product-mix will necessitate larger investments. It would be desirable to phase out the concessional rates of interest as recommended recently by the Committee on the Financial System (GoI 1991), because the normal rates of interest charged by the institutional sources would still be lower than those charged in the informal market. It would, however, be necessary to ensure the timely availability of institutional credit to the weaker sections through appropriate rationing of credit. Otherwise, the well known tendency on the part of the bankers to reduce unit costs by serving a smaller number of big borrowers rather than a large number of small borrowers would result in squeezing credit to the needy sections. Incidentally, the recommendation of the Narasimham Committee that the credit share for the 'redefined priority sector' consisting of weaker sections be pegged at the existing level of 10 per cent does not take into account the need to replace the informal sources on which these sections are heavily dependent at present, let alone meeting the growing demand for credit from them. However, this particular recommendation of the Committee was not accepted by the Reserve Bank of India. Subject to certain broad guidelines, the managements of credit institutions have to be accorded greater autonomy by doing away with administrative and political interference in their day-to-day functioning.

Decentralisation

The designing and management of infrastructure such as surface irrigation and rural electrification is done by the top-down bureaucratic structures with very little say for the concerns and perceptions of user-farmers. The absence of a link between the payment of service charges and the prospect of improvement in the infrastructural services has provided a fertile ground for the politicisation of cost-recovery processes. Involvement of the user-farmers in the management of agricultural infrastructure through the users' associations and the local level elected institutions represents the closest approximation to privatisation of such activities. Such non-market institutions of representative type at the local level including the voluntary organisations seem to provide the only workable alternative at this stage, whether to bureaucratic management or to full-scale privatisation. The recent agricultural breakthrough achieved in West Bengal points to the efficacy of decentralised management of infrastructure (Sen 1993).

The political and bureaucratic resistance to decentralisation notwithstanding, the need as well as the prospects for decentralisation are much greater now than ever before. The Constitution (Seventy-third) Amendment Act, 1993, has given some impetus to the processes of decentralisation insofar as it mandates periodic elections to the *panchayati raj* institutions, reservation of seats for scheduled castes and scheduled tribes and for women. However, the states have to muster the necessary political will to delegate to these institutions, the vital subjects bearing on economic development and social justice as listed in the Eleventh Schedule; to devolve adequate resources to them through the mechanism of the state-level Finance Commission; and to entrust them with necessary power over the administration. The realisation of these goals is bound to be a painstaking and time-consuming process.

The success of decentralisation in achieving its social objectives would very much depend upon the prevailing social structure. It is not accidental that *panchayati raj* institutions have been functioning well in situations where land reforms have been implemented effectively. Because of low wages and scale-neutrality of bio-chemical technology, there is, as yet, no evidence of increasing returns to scale in Indian agriculture. There is, therefore, no case on efficiency grounds for doing away with the existing ceiling on landholdings and for encouraging large-scale corporate farming at this stage, except in cases where land, hitherto uncultivated, cannot be reclaimed without large-scale investments in heavy machinery and equipment. It would, however, be necessary to remove restrictions on the lease market for agricultural land with a view to ensuring efficient use of resources as well as for augmenting the operational holdings of the small and the marginal farmers.

Trade Liberalisation

As mentioned in Chapter 1, there is considerable potential for raising India's share in world exports of agricultural commodities. It is necessary to exploit this potential in order to widen the opportunities for raising farm income and employment by diversifying and modernising agriculture. Dryland areas where wages are low and poverty is widespread may benefit significantly from trade, as they have a comparative advantage in horticulture and livestock products, the demand for which is less inelastic than for foodgrains. However, this would require large investments in agro-processing and other infrastructure.

Conclusions

Of the six sets of reforms discussed above, freeing trade for 'setting prices right' is perhaps the easiest to achieve. But ensuring adequate supply response and equitable distribution of gains depends critically on the non-price factors like stepping up investment, technological change, conservation of land and water resources, credit reform and decentralisation of management which are far more difficult to achieve. Therefore, 'freeing' trade without simultaneously undertaking measures for augmenting capacities and building 'safety nets' for the poor may result in slow growth, high prices of foodgrains and the accentuation of regional disparities in development. Excessive pre-occupation with prices may prove to be harmful by diverting attention from the more important tasks which belong to the non-price world for releasing constraints on production such as investment and technological change (Narain 1976). The experience with agricultural supply response in several developing countries shows that the elasticity of output with respect to major technological shifters like irrigation is significantly higher than the price elasticity even after the effect of price-induced innovations is accounted for (Krishna 1982). "National Policymakers and international development bankers will therefore do well to devote at least as much attention and effort to the development of technology, infrastructure, and human capital as to the price environment" (Krishna 1982).

References

Government of India, Planning Commission (1992). *Eighth Five-Year Plan*, Vol.II, New Delhi.

Gulati, Ashok (1990). "Fertiliser Subsidy: Is the Cultivator 'Net Subsidised?'," *Indian Journal of Agricultural Economics*, Vol.XVL, No. I, January-March.

Jain, H.K. (1993). *The GATT Proposals on Agriculture: Their Scientific Implications* (unpublished), National Institute of Science, Technology and Development Studies, New Delhi.

Krishna, Raj (1982). "Some Aspects of Agricultural Growth, Price Policy and Equity in Developing Countries," Food Research Institute Studies, Vol. XVIII, No. 3.

Narain, Dharm (1976). "Growth of Productivity in Indian Agriculture" in Raj, K.N., Amartya Sen and C.H. Hanumantha Rao (eds.), *Dharm Narain - Studies in Indian Agriculture*, Oxford University Press, Delhi, 1988.

Narasimham, M. (1993). *Economic Reforms: The Unfinished Agenda*, Rajiv Gandhi Institute of Contemporary Studies, New Delhi.

———. (Chairman) (1991). *Report of the Committee on the Financial System*, New Delhi.

Rao, C.H.H. (1994). *Agricultural Growth, Rural Poverty and Environmental Degradation in India*, Oxford University Press, New Delhi.

Reserve Bank of India (1989). *A Review of Agricultural Credit System in India: Report of the Agricultural Credit Review Committee*, Bombay (Chairman: A.M. Khusro).

Sen, Abhijit (1993). "Agriculture in Structural Adjustment," paper presented at the seminar on *Agricultural Reform in India*, Indira Gandhi Institute of Development Research, January 11-12.

Thamarajakshi, R. (1977). "Role of Price Incentives in Striulating Agricultural Production in a Developing Economy" in Douglas Ensminger (ed), *Food Enough or Starvation for Millions*, Tata McGraw-Hill, New Delhi.

This paper is based on the Dharm Narain Memorial Lecture delivered by the author on January 17, 1994 under the joint auspices of the Institute of Economic Growth and the Delhi School of Economics, University of Delhi; Published in *Economic and Political Weekly*, Vol. XXIX, Nos. 16 and 17, April 16-23, 1994.

17

Economic Reforms, Agriculture and Rural Development

The economic reforms pursued in the country so far have barely touched agriculture and rural development. While the idea of releasing market forces is gaining ground under the impact of the on-going economic reforms, we are nowhere near the idea of broad-based participatory rural development.

Our experience with economic reforms has been encouraging so far, especially when compared to the experience with reforms in several countries of Latin America and Sub-Saharan Africa. The growth rate of GDP has recovered from the dip in the immediate post-reform period. The economy now seems to be poised for a sustained growth rate of around 7 per cent per annum. Incidentally, this is the target set for the Ninth Five Year Plan. But, is this growth in GDP percolating adequately to the poor and the underprivileged sections in rural areas? This does not seem to be so going by the evidence on agricultural growth in general and the growth in foodgrains in particular.

There are clear signs of a slow down in agricultural growth in the post-reform period when compared to the eighties. For example, according to the Economic Survey (1996-97), foodgrains output in the nineties has been growing at a rate less than the population growth rate. Real public investment in agriculture continues to be lower than that realised in the early eighties. As a result, there has been a significant shortfall in the achievement of Eighth Plan's irrigation target, especially from the major and medium irrigation projects where the achievement is less than half the target. The growth rate in the consumption of fertilisers has been less than half of that achieved in the eighties. There has been a steep fall in the share of credit from the commercial banks allocated to agriculture, despite a significant improvement in the recovery of their loans advanced to farmers. This shows that the country has not been able to fully exploit the potential offered by a series of good monsoons in the recent period. In the light of this experience, the target of 4.5 per cent growth rate set for agriculture over the Ninth Plan, though desirable, seems difficult to achieve unless a credible package of measures is put in place.

The experience with economic reforms in several countries has demonstrated substantial benefit flowing to agriculture. The exchange rate liberalisation and reduction in the rate of protection to industry get translated into improved terms of trade for agriculture. These indirect benefits for agriculture are far more important than the direct benefits from opening up agriculture and doing away with various controls and regulations. Thus the incentive framework for agriculture would improve vastly and new opportunities would be opened up. But agriculture can respond to these opportunities favourably only if the supply constraints are eased. It is even more difficult to ensure that agricultural growth so achieved would be broad-based and employment-oriented so that the growth realised results in a significant reduction in poverty.

The experience of East and South-East Asian countries is extremely relevant for us in this regard. In these countries agricultural growth has been quite high even before they embarked upon economic reforms. The levels of infrastructure development there, such as irrigation, rural roads and markets were already quite high before the start of economic reforms. The levels of development in respect of primary education and health and institutional framework like land tenures and systems for the delivery of credit and inputs were far more conducive to wider sharing of benefits than in our case.

In several of these countries the growth of GDP originating from agriculture was well over 1 per cent or close to 2 per cent per head in the seventies and the eighties. On the other hand, in most of South Asia, including India, the per head GDP growth originating from agriculture was negative in the seventies. The performance improved in the eighties, but the per head growth was still below 1 per cent in most cases. In countries like China, Indonesia and Thailand, the post-reform growth was spearheaded by agriculture resulting in the speedy alleviation of rural poverty. The experience of East and South-East Asia shows that we in India need to aim at 4 per cent to 4.5 per cent agricultural growth, or 2 per cent to 2.5 per cent per head, together with the overall GDP growth rate of 7 per cent, if we are serious about the speedy alleviation of poverty.

At this stipulated level of development, agriculture would get diversified and its growth would increasingly consist of dairying and animal husbandry, fisheries, horticulture, and floriculture, etc. With this composition of output, agro-processing industries in rural areas will receive a special impetus meeting the domestic as well as export demand. Such activities will be labour-intensive and will have strong linkages with manufacturing and services sectors. It is not surprising, therefore, that despite a significant decline in the share of agriculture in GDP in several of the East and South-East Asian countries, the proportion of labour force engaged in the rural sector including in agriculture is still quite high, exceeding 60 per cent of total labour force in countries like China and Thailand. It is this broad-based labour-intensive rural development together with high levels of social development such as primary education and primary health services which has triggered off a speedy demographic transition in many of these countries. Apart from leading to high rates of savings and investment, this process has activated the trickle-down mechanisms, so that high growth resulted in the speedy alleviation of poverty.

The structural adjustment measures represented by the deregulation and liberalisation of the economy, underway in India, have no doubt created the necessary incentive framework for agriculture and for a broad-based labour-intensive rural development, the beneficial impact of which will be felt over a long period. Already, private investment in agriculture has picked up considerably, not only recovering from the down trend of the eighties but recording in the last year the highest level of real investment ever achieved. However, the economic reforms implemented so far have failed to ensure the necessary physical and institutional infrastructure so as to evoke sufficient supply response from agriculture.

The manner in which the stabilisation measures have been pushed through have adversely affected agriculture and rural development. The target for reducing fiscal deficit has been pursued not by raising revenues or by reducing inessential expenditure but mainly by reducing capital expenditures. The tax ratio or the tax revenues (net to the Centre) as a proportion to GDP is still lower by about 1 percentage point when compared to the pre-reform level. The hidden subsidies at the state level represented by the non-recovery of irrigation charges and power tariff continue to be as high as in the pre-reform period. What is more, certain state governments seem to be totally unaffected by the post-reform ethos and are going ahead with increased subsidies and even "free" inputs to farmers.

The major casualty of this approach is public investment in agriculture at the state as well as the central level. Although private investment in agriculture has been picking up, of late, under the impact of economic reforms, it is unlikely to enter the infrastructure sector in any significant way in the immediate future, which has to be provided through public investment, especially in the less developed areas. The combined outlay of the Centre and States on rural development and social services as a ratio to GDP does not seem to have been restored even to pre-reform level. There has no doubt, been a justification for a gradual rise in the farm gate prices of fertilisers concurrently with the rise in the procurement prices of crops. But a sudden and steep rise in the prices of fertilisers in the wake of economic reforms following a reduction in subsidy led to a significant decline in the demand for fertilisers. On account of the resulting decline in output and income, a large number of small and marginal farmers were pushed below the poverty line in the immediate post-reform period. The decontrol of phosphatic and potassic fertilisers and the consequent doubling of prices has accentuated the existing imbalance in the application of different nutrients resulting in the decline in the productivity of fertilisers as a whole.

In order to enable the rural sector to respond to the challenges in the post-reform era, a number of reforms have to be carried out in the areas of resource mobilisation, public expenditure, management of infrastructure and rural development programmes. The economic reforms undertaken so far, though momentous in their consequences, were the easier ones which could be implemented with a stroke of pen, so to say, as they encountered little resistance from the vested interests. The reforms ahead, on the other hand, are the most difficult ones to implement. Stepping up investment in infrastructure through cost-recovery is inextricably linked with the quality of services provided and how the infrastructure and other services are managed. It is naive to expect that the farmers in a democratic polity would be willing to pay for the services regardless of the quality and cost of such services, especially when competitive populism is rampant. If there were any illusions on this score among the intellectuals and policymakers, these would have been shattered by now.

The management of our rural infrastructure, *e.g.*, irrigation, electricity and credit is highly centralised, bureaucratised and politicised. This indeed accounts for the low quality of services, their high cost and populist overtones which together result in the failure to recover costs. The debate on this issue has been focused on cost-recovery, as it has an immediate bearing on fresh investment to expand services. However, deficiencies in management and pricing have had other adverse consequences such as the wasteful use of scarce inputs, degradation of environment on account of water-logging on the one hand, and depletion of ground water, on the other. These have also led to inequalities as between different regions and between different sections of farmers. Decentralised management of rural infrastructure through de-bureaucratisation and de-politicisation should, therefore, be top on the agenda of economic reform.

In the first place, the associations of user-farmers and non-governmental organisations could be entrusted with the responsibility for repairing and managing small irrigation works with the power to collect water rates and retain a part of such revenues for maintenance and development. Secondly, the local elected institutions at *Panchayat* and higher levels could be entrusted with similar functions including the transmission and distribution of power with the authority to levy and collect the service charges. Thirdly, investment in such infrastructure by the private sector, including subscription of shares by the farmers on a wide scale, could contribute a great deal towards augmenting and efficiently managing the infrastructure.

The experience with decentralisation of such services in South-East Asia has shown that irrigated yields can be increased by at least 50 per cent and the farmers would be willing to pay service charges to the extent of at least one-fourth of the increment in the value of output, as the benefits to them from better maintenance of works are visible. Such payments by the farmers in India would represent a nearly ten-fold increase over the existing levels. It is heartening that some state governments like Gujarat, Karnataka and Andhra Pradesh have already taken certain steps in this direction.

Higher outlays on social sectors are necessary but unless the schemes are devised and implemented by the Panchayati Raj institutions, large leakages my persist and the benefits may not reach the target groups. The 73rd amendment of the Constitution was designed precisely to plug such loopholes. Whereas periodic elections and reservations to the weaker sections and women in these bodies have become mandatory, most state governments failed so far to muster enough political will to devolve the necessary functions, resources and power over administration to these bodies. However, without effective decentralisation of power, economic reforms will not yield the desired benefits.

Convocation address delivered at the Sixteenth Convocation of the Institute of Rural Management, Anand, Gujarat, April 10, 1997.

II

Farm Size and Tenure

18

Farm Size and the Economies of Scale

It is argued here that unlike in manufacturing industry, the technical conditions of crop production do not impose the kind of constraints which make the enlargement of farm size indispensable for absorbing improvements in farming techniques or for the optimum utilisation of indivisible equipment. The available empirical evidence for India suggests that the costs on account of indivisible factors such as bullocks cannot be considered to be particularly higher on smaller farms where family labour is abundant and that, owing to the labour-intensive techniques of cultivation, managerial and supervisory bottlenecks are conspicuous on larger farms. These factors, together with the fact that under conditions of family-farming the farm production unit is intertwined with the family consumption unit, explain why small farms are invariably characterised by higher capital value of land and larger income per acre.

According to the theory of the firm the internal economies of large scale arise upto a point owing to (a) specialisation and division of labour in the production processes made possible by technical progress; and (b) optimum utilisation of indivisible equipment as well as management. Underlying these advantages are certain technical constraints common to manufacturing industries under conditions of factory production. Technical progress and division of labour tend to make the size of the plant bigger. Most of the machinery of the plant is technically immobile so that the optimum utilisation of such equipment necessarily entails the expansion of output.

Agriculture, on the other hand, is essentially a biological process, land being the primary agent. Due to the fixed nature of this biological process in regard to the nature of the commodities to be produced, the production period and the technical processes, the scope for specialisation and division of labour is very limited. Technical progress, to the extent possible, assumes the forms of external aids to land such as better tilling, improved seeds, fertilisers, insecticides, *etc*. Since land as a factor of production is perfectly divisible and since many inputs, *e.g.* labour, water, seeds, manures and insecticides, *etc*., are equally divisible, improved methods can be applied to land, irrespective of the latter's size.

Optimum Utilisation Possible

Insofar as equipment, like a pair of bullocks or a machine, is not technically tied up with land and can and has to be removed from land once its operations are over, the optimum utilisation of this mobile, though indivisible, equipment is possible by employing it on different farms on a hire or exchange basis. Despite this technical possibility, some element of under-utilisation of such equipment can be expected owing to institutional and psychological obstacles to adjustment, especially under conditions of seasonal demand for the equipment.

In the case of management, however, the economies of scale can be realised upto a point, especially in the long-run. Since farming processes require close personal attention at each stage and since the scope for division of functions and the delegation of responsibility is very limited, managerial and supervisory bottlenecks arise at comparatively lower levels of farm size. This stage will be reached much earlier under labour-intensive techniques of cultivation, as the human problems of management can be expected to be particularly acute in such a situation.

Under the system of family-farming such as in India, the owners or cultivators are themselves the managers who also perform manual work in most cases. The existing mode of cultivation does not require highly specialised technical skills on the part of these managers. Nor have the entrepreneurial abilities a very significant role to play especially as farming is carried on partly for household consumption. Supervision of farm work is really the most important part of managerial functions under the existing conditions. Viewed thus, management is not a very scarce input among small farms where there is an abundance of family labour. Also, the prevailing labour-intensive techniques of cultivation render managerial or supervisory functions very difficult to perform on large farms. Therefore, from the standpoint of efficiency, the problem is not so much the under-utilisation of management input on smaller farms as the diseconomies of large scale operations arising from managerial and supervisory bottlenecks on larger farms.

Bullocks and implements thus constitute the major indivisible factor in Indian farming. The Farm Management Studies have shown that a large number of cultivators in several parts of the country, who belong to the small size groups do not own any plough-cattle and that those owned by others are better utilised due partly to the larger cultivated area per pair of bullocks on the larger farms and partly to their being used on hire or exchange basis by smaller farmers. In Madhya Pradesh, for instance, about 28 per cent of the cultivators, most of them cultivating less than five acres, do not own any bullocks. About 7.5 per cent of the farmers own one bullock only, while 38 per cent have a pair of bullocks each (Madhya Pradesh Report, 1955-56, p. 24). Similarly in Andhra, as many as 40 per cent of cultivators, most of them belonging to small size groups, do not own draught cattle; 29 per cent own one pair of bullocks and only about 13 per cent have two pairs each. (Andhra Report, 1957-58, p. 160). In West Bengal, 35 per cent of holdings have no draught cattle and 6.1 per cent have only one bullock each (West Bengal Report, 1954-55, p. 12). A similar tendency, though not equally pronounced, has been noticed in Madras and Bombay. In Punjab, the small farmer tries to minimise costs on this item by maintaining bullocks of inferior quality which cost less. The larger holdings, on the other hand, maintain bullocks of better quality which are better fed, with the result that the cost on a pair of bullocks increases with increase in the size of holding (Punjab Report, 1954-55, pp. 54-57).

The market prices of bullocks cannot be taken as reflecting the true costs to the farmers. For one thing, a significant proportion of these bullocks are farm-bred, family land and family labour being the main constituents of costs on this item. The practice of valuing farm-bred animals on the basis of market prices for costing purposes without, at the same time, including these values on the output side can hardly be justified. Also, the existing method of rating family labour spent on the upkeep of animals at market wage rates serves to inflate the bullock costs among small size-groups. It is interesting to note in this connection that according to the present methods of cost calculation, bullock costs as a ratio of gross output are higher wherever the proportion of family labour and family land is higher (Table 18.1). In West Bengal, Andhra and Madhya Pradesh where, as noticed above, hire and exchange arrangements for sharing bullocks are common, the proportion of bullock costs to output is very low. Again, it is in these regions that the proportion of family labour to total labour is particularly low. The proportion of family land is also relatively small. It is possible that in view of higher costs on hired labour and rented land, the cultivators in these regions try to minimise the costs on this indivisible equipment. In Uttar Pradesh, Madras and Bombay, where the costs on bullocks are quite high and the practice of sharing them is less common, the bullocks are presumably under-utilised. But the proportions of family labour and family land are quite high. The readiness on the part of the cultivators to put up with under-utilisation of bullocks in these regions may be explained, to a certain extent, by the relatively low opportunity costs of their family inputs used in maintaining the bullocks.

Table 18.1

Costs on Indivisible Equipment, the Proportion of Family Land and Family Labour

Region	*Costs on Indivisible Equipment (Percentage of Gross Output)*		*Family Land (Percentage of total)*	*Family Labour (Percentage of total)*
	Bullocks	*Implements*		
	(1)	*(2)*	*(3)*	*(4)*
Uttar Pradesh	28.9	3.9	100.0	71.0
Punjab	15.2	3.0	65.0	77.0
West Bengal	6.6	1.1	70.7	64.0
Andhra Pradesh	8.4	2.1	74.0	25.0
Madhya Pradesh	11.3	2.1	70.0	24.2
Madras	17.2	3.9	79.3	75.0
Bombay	18.6	4.0	81.2	75.4

Sources: For columns (1) and (2), the same as given below Table 18.2. For columns (3) and (4) "Studies in Economics of Farm Management"; Uttar Pradesh, 1954-55, p. 37; Punjab, 1954-55, p. 31 and 71; West Bengal, 1954-55, Appendix Tables, pp. 30, 45 and 46; Andhra Pradesh, 1957-58, pp. 60 and 216; Madhya Pradesh, 1955-56, pp. 17 and 158; Madras, 1954-55, pp. 30 and 63; and Bombay, 1956-57, pp. 22-23, 1954-55, p. 218.

Comparing Costs and Returns

Size-efficiency analysis necessarily involves some appropriate method of comparing costs and returns. Insofar as the opportunity costs of family-labour cannot be regarded as approximating the ruling wage rates, the estimates of total costs based on the imputed values would give an unrealistic and even misleading picture. This is evident from the findings of the Farm Management Studies mentioned above. On the basis of total costs (or cost C) which includes also the imputed values, farm-business is found to be unremunerative. Alternatively, since the true price of family labour cannot be regarded so low as approximating zero, 'cost' consisting of only paid-out items would yield a distorted picture. For, in such a case, the differences in cost-return ratios would reflect not only the differing 'efficiency' but also differences due to inter-farm variations in the proportions of family labour in the total labour input. Similarly, the cost-return ratios may be distorted due to differences in the proportions of family and rented land. Since the community is interested in maximising the net income or net value added from agriculture consisting of the returns on land, labour and capital, whether owned or hired, a meaningful concept of cost would be the one consisting of the inputs to be deducted from gross output for arriving at the above net income. Since land constitutes the single biggest item in total costs and since it is very scarce relative to labour, net income per unit of this scarce factor could serve as a meaningful, though partial, indicator of farm efficiency for economic policy.

Table 18.2 gives the figures of average net value added or net-income per acre per annum for various size-groups in different regions for which data from Farm Management Studies are available. Net income is obtained by deducting from the gross value of output the input values of seeds, manures, fertilisers, irrigation charges, feed of bullocks and the depreciation on fixed equipment, *e.g.*, bullocks, implements, machinery and farm constructions. Wherever possible, labour charges (both family and hired) for the upkeep of animals have been separated from bullock costs. A significant fact emerging from Table 18.2 is that net income per acre is relatively higher among small size groups and tends to decline among the larger farms. Land is expressed here in terms of simple acre, unstandardised with respect to fertility. But there is no reason to think that the fertility of soil would be lower on larger farms.

Table 18.2

Net Income and Rental Value

(In Rupees Per Acre)

Size-group (acre)	*Uttar Pradesh*	*Punjab*	*West Bengal*	*Andhra Pradesh*	*Madhya Pradesh*	*Madras*	*Bombay*
	(1)	*(2)*	*(3)*	*(4)*	*(5)*	*(6)*	*(7)*
Upto 1.25			184	354 (204)		132	
1.26 - 2.50			188	305		(50)	
2.51 - 3.75	126	130 (73)	184	(192)	79		54 (18)
3.76 - 5.00			162	285 (175)		156 (50)	
5.01 - 7.50	137	129 (67)	200	236 (168)	69	126 (39)	70 (11)
7.51 - 10.00			195	234 (176)		129 (43)	
10.01 - 15.00	117	119 (59)	136	213 (169)	71	89 (27)	36 (12)
15.01 - 20.00	129			282 (185)	66	45 (21)	36 (13)
20.01 - 25.00					66	61 (30)	34 (8)
25.01 - 30.00							43 (7)
30.01 - 40.00	110	104 (55)	163	239 (151)	67	44 (15)	37 (7)
40.01 - 50.00		109 (49)			71		29 (6)
Above 50.00					67		

Note : The figures of rental values are given in brackets. Size-wise figures of rental values of owned land are not available for Uttar Pradesh, West Bengal and Madhya Pradesh.

Sources: Studies in Economics of Farm Management: Uttar Pradesh, Punjab, West Bengal, Madras and Bombay, for the years 1954-55, 1955-56 and 1956-57; Madhya Pradesh for the years 1955-56 and 1956-57; and Andhra Pradesh for the years 1957-58 and 1958-59.

The declining tendency does exist in regard to the current fertility of land which is made up of the original fertility and fertility resulting from man-made improvements on land — the latter indicating the efficiency of land utilisation. The average per acre rental values for each size of holding for some regions given in brackets clearly indicate that the capital values per unit of land must be higher among smaller size groups and that these values decline as the farm size increases. These higher values reflect in part the availability of irrigation which is proportionately higher on smaller size farms (see Farm Management Studies). This itself is an indicator of a higher degree of capital formation. Partly, these differences reflect variations in other man-made improvements on land which also represent capital formation. It is reasonable to suppose that the lower the level of income of the cultivator, the greater will be his attachment and concern for his land — the principal means of production — and consequently the greater will be the incentive to make improvements on it.

'Smallness' no Deterrent

These facts indicate that the 'smallness' of size has not proved a particular deterrent to capital construction through the application of surplus labour. Nor has it prevented the cultivators from realising higher net income per acre. If the element of over-estimation in the present figures of bullock costs for small size groups is eliminated, the net incomes per acre would turn out to be even higher. At the same time, these figures cannot be taken to imply that the higher net-income per acre among smaller size-groups is attributable to the 'smallness' of size *as such*. Rather, they are traceable to factors making for intensity of cultivation, *viz*, better irrigation, intensity of cropping and larger application of family labour. It is obvious that upto a point in the size range these factors have little to do with farm-size as such but are prompted by the need to secure a minimum subsistence for the farm-family. The efficiency of the managerial factor among these size-groups may be a contributing factor. Among the large-sized farms, on the other hand, management may be both insufficient to meet the increased needs of supervision and inefficient owing to the dissociation of ownership from management on the one hand and of management from work, on the other. In fact, this may explain why labour input per acre among large size-groups is found to be relatively lower (see Farm Management Studies).

Thus, from the standpoint of efficiency the basic problem is not so much the 'uneconomic' nature of small holdings as the insufficiency of inputs, particularly labour, and the consequent low productivity per acre among larger farms — particularly those above 10 acre which are known to account for over two-thirds of total cultivated area. The above analysis suggests that under the system of family-farming and labour-intensive techniques, productivity of land and total output can be increased if the structure of land-ownership is altered and a more even distribution is ensured as this would, among other things, make for greater identity between the factors of ownership, management and labour. Alternatively, if the existing structure of land ownership is not to be disturbed, then the introduction of mechanised processes would become necessary for a more effective utilisation of land, as this would lessen the supervisory and managerial bottlenecks on larger farms. Of these two alternatives, the former has already proved far from politically feasible. The latter, though not quite desirable, at any rate in the immediate future from the economic point of view, is being gradually adopted in many parts of the country. In view of the existing high man-land ratio and the concentration of land-ownership, the mobility of land, to the extent it takes place from the larger to the smaller or middle-sized groups through sales and purchases and through tenancy on a commercial basis, should be considered desirable.

The Economic Weekly, December 14, 1963.

19

Alternative Explanations of the Inverse Relationship between Farm Size and Output Per Acre in India

Introduction

The Farm Management Studies conduced by the Ministry of Food and Agriculture in certain "typical" regions of India during mid-fifties revealed that output per acre declined with the increase in the size of holding.[1] Several explanations have been offered for this inverse relationship between farm size and output per acre.[2] It has been suggested that in so far as the inferences drawn in Farm Management Studies are based on the data grouped according to size-classes of holdings these cannot be considered as well-established unless tested on the basis of individual farm data.[3] An attempt has been made in this paper to study the behaviour of inputs and output per acre with increase in farm size and to test some of the current hypotheses concerning productivity behaviour on the basis of individual farm data relating to Bombay and Andhra.[4] The limitations of some of the current explanations as well as the relevance of certain other explanations are brought out on the basis of the available empirical evidence.

Output and Inputs Per Acre According to Farm Size

The relationship between farm size and output per acre may be measured by fitting an exponential function of the type

$Y=aX^b$ to the individual farm data where Y may be taken either as output per acre or as total output of the farm. If 'Y' is output per acre, the regression coefficient 'b' would have a negative sign if output per acre declines with the size of holding; if Y is total output, b would be positive but less than unity indicating that output increases less than proportionately to the farm size which means a decline in output per acre. It has been widely observed that just as the use of grouped data for a linear function introduces aggregation bias and exaggerates the value of R^2 (the per cent variation in the dependent variable explained by the independent variables), the use of data on a per acre basis introduces an element of non-linearity. For the functions fitted in this study, therefore, the second method is followed *i.e.*, using the data at the individual farm level on a non-per acre basis. The results are as follows:

$$\text{Log } Y = 1.95 + \underset{(0.01)}{0.74} \log X_1 \quad \text{(Bombay: Ahmednagar)}$$

$R^2 = 0.43$; $N = 80$

$$\text{Log } Y = 2.29 + \underset{(0.01)}{0.55} \log X_1 \quad \text{(Bombay: Nasik)}$$

$R^2 = 0.29$; $N = 80$

$$\text{Log } Y = 2.43 + \underset{(0.04)}{0.74} \text{ Log } X_1 \quad \text{(Andhra: Karimnagar)}$$

$R^2 = 0.92$; $N = 33$

Where Y = Farm output in rupees; X_1= Farm size in acres; R^2 stands for the variation in the dependent variable Y explained by the independent variable X_1 and N for the number of observations.

The regression coefficients with respect to farm size are positive and less than one. They are significant at 1 per cent level and are significantly different from unity indicating diminishing returns to land size. These results agree with the findings of the Farm Management Studies arrived at on the basis of the grouped data that output per acre declines with increase in the size of holding.

As is evident from the Farm Management Studies, this decline in output per acre with the size of holding is traceable to a corresponding decline in the inputs per acre. This is confirmed by the following equations:

$$\text{Log } C = 2.25 + \underset{(0.01)}{0.55} \log X_1 \text{ (Bombay: Ahmednagar)}$$

$R^2 = 0.33$; $N = 80$

$$\text{Log } C = 2.37 + \underset{(0.01)}{0.48} \log X_1 \text{ (Bombay: Nasik)}$$

$R^2 = 0.40$; $N = 80$

$$\text{Log } X_2 = 2.40 + \underset{(0.05)}{0.72} \log X_1 \text{ (Andhra: Karimnagar)}$$

$R^2 = 0.85$; $N = 33$

Where C = Total cost including the imputed value of family labour and rental value of own land; X_1= Farm size in acres; X_2= Labour and capital inputs per farm in rupees. All the coefficients are significant at 1 per cent level and are significantly different from unity.

According to the Farm Management Studies, the proportion of area irrigated invariably declines with the size of holding. To know the effect of irrigation on output per acre according to the size of holding, the following equations are derived by including the proportion of area irrigated (I) as an independent variable along with size of holding:

$$\log Y = 1.55 + \underset{(0.08)}{0.87} \log X_1 + \underset{(0.14)}{0.80} \log I \text{ (Bombay: Ahmednagar)}$$

$R^2 = 0.61$; $N = 80$

$$\log Y = 1.96 + \underset{(0.08)}{0.65} \log X_1 + \underset{(0.12)}{0.80} \log I \text{ (Bombay: Nasik)}$$

$R^2 = 0.55$; $N = 80$

$$\text{Log } Y = 2.09 + \underset{(0.06)}{0.81} \log X_1 + \underset{(0.13)}{0.19} \log I \quad \text{(Andhra: Karimnagar)}$$

$R^2 = 0.94$; $N = 33$

These results show that as compared to the first set of equations, where output is related to farm size only, the per cent variation in output explained by farm size after correcting for irrigation is invariably higher. Also, the coefficients with respect to size are raised in all cases, indicating that the decline in output per acre with size becomes less marked after correcting for irrigation.

Labour-Based Explanation

Why do inputs vary less than proportionately to the size of holding? A.K. Sen offered an interesting explanation in terms of the low opportunity cost of family labour to small farmers.[5] Sen's argument would imply that the excess of labour input per acre among smaller farms over larger ones belongs to a range where marginal net product of labour is less than the ruling wage rate. Since, as noted above, the complementary inputs such as capital and irrigation also vary less than proportionately to size, the marginal net product curve of labour cannot be identical for varying sizes. Owing to the larger capital input per acre (capital being of labour-absorbing type) and greater managerial efficiency among the small farms, the marginal net product curve of labour would be higher than the corresponding curve for the larger farms and, even if further employment of labour is stopped at the point where the marginal net product is equal to wage rate, the labour input and output per acre on smaller farms would be higher. It is irrelevant to this argument whether the capital among smaller farms is constructed through the use of family labour or through monetary investments. Indeed, the very possibility of capital construction through the application of family labour would cause the marginal net product of labour to be higher than what it would otherwise be.

However, it is possible that family labour and capital built by family labour are stretched beyond the remunerative level (in the market sense) among smaller farms even with a higher marginal net product curve. Sen's argument based on 'costless labour' would apply for such a situation.

Although all family labour is costless in the sense that it is not paid for, we are concerned here only with that part of family labour which is applied beyond the profitable range. For the range where the marginal net product of labour is higher than or equal to wage rate, the cheap-family-labour argument fails to explain larger labour input and output per acre among smaller farms because it is always open to larger farms who have a greater command over capital, unless otherwise constrained (to which we shall come later), to effect a similar shift in the marginal net product curve of labour through the provision of more irrigation and other improvements.

Following his labour-based explanation Sen explains that the "losses" in farm business shown to have been incurred by the small farmers and the increase in profit per acre with the size of holding are attributable only to the imputation of wage rates to the family labour.[6] Sen clearly implies here that the decline in the proportion of family labour inputs *per se* explains the decline in the element of "loss" due to the imputation of wage rate. Over a considerable range in farm size where the need to hire permanent farm labour arises, *i.e.* where hired labour is used all the year round, the application of family labour could be within a profitable range because some labour is hired at margin throughout the year and a mere decline in the proportion of family labour input cannot explain the increase in profit per acre. It is shown later that the necessity to hire some labour at margin arises throughout the year for farms of above 10 to 15 acres and that even these farms show a decline in output as well as an increase in profit per acre with increase in size. Therefore, the explanation for the increase in profit per acre might lie elsewhere. The Farm Management Studies show that there is considerable under-utilisation of indivisible capital, *e.g.* bullocks and implements, among smaller farms.[7] It is a matter for investigation how profit (or loss) per acre would behave with size if indivisibility factor is held constant. But *a priori* it would appear that the losses among smaller farms or the rise in profit per acre with size can be explained to a significant extent by the existence of indivisibilities.

Fertility-Based Explanation

Of all the explanations offered by Sen, fertility-based explanation[8] seems to be the weakest. By fertility Sen means natural or original fertility of soil as distinct from its current productive capacity resulting from improvements. According to him output per acre may show a decline if natural fertility of soil declines with size. He then offers an explanation for the inverse relationship between size and fertility: those who are originally settled on fertile soils would multiply faster owing to the larger per capita income and this would lead to greater sub-division of such holdings. An inverse correlation between size and fertility would thus be established. But any attempt to explain a phenomenon should be proceeded by a reasonable assurance as to its factual basis.

Khusro has shown that land revenue per acre declines with size but he does not claim that it is an index of the declining natural fertility of soil with size.[9] This is understandable because land revenue ratings reflect, among other things, improvements like irrigation which show an inverse relationship to size. Khusro found that the decline in output per acre with size becomes less marked when the holdings are standardised on the basis of land revenue ratings. This, again, is what one would expect when the productive capacity of land is considered relatively homogenous among different holdings. The available evidence suggests that the decline in the rental value of land (which reflects the current productive capacity) with size is substantially explained by the decline in the irrigation factor. That the rental value of holding varies less than proportionately with size is indicated by the following equations:

$$\log Y = 1.26 + \underset{(0.01)}{0.78} \log X_1 \text{ (Bombay: Ahmednagar)}$$

$R^2 = 0.53$; $N = 80$

$$\log Y = 1.53 + \underset{(0.01)}{0.58} \log X_1 \text{ (Bombay: Nasik)}$$

$R^2 = 0.29$; $N = 80$

Where Y = rental value in rupees per farm; and X_1= farm size in acres. Both the coefficients are statistically significant and are significantly different from unity.

The inclusion of the proportion of holding irrigated as another independent variable gave the following results:

$$\log Y = 0.80 + \underset{(0.06)}{0.94} \log X_1 + \underset{(0.10)}{0.93} \log I \text{ (Bombay: Ahmednagar)}$$

$R^2 = 0.78$; $N = 80$

$$\log Y = 1.23 + \underset{(0.09)}{0.66} \log X_1 + \underset{(0.13)}{0.74} \log I \text{ (Bombay: Nasik)}$$

$R^2 = 0.56$; $N = 80$

The coefficients for land size are raised significantly with the inclusion of irrigation factor. In the case of Ahmednagar the coefficient is not significantly different from unity which means that rental value bears more or less a proportionate relationship with the farm size after correcting for the irrigation factor. If improvements other than irrigation vary inversely with size, the tendency revealed by the above equations would be reinforced if such improvements can be included as an explicit variable. Since lands of today owe their fertility "as much to nurture as to Nature",[10] any attempt to disentagle them with a view to knowing the behaviour of each with size must remain a wild-goose chase. One might however, argue, deriving support from history as well as existing facts, that those who had to settle down originally on small holdings, because of lower income, have been making improvements on land over several generations through the application of family labour and this has resulted in an inverse correlation between farm size and productive capacity of land.

Tenancy-Based Explanation

If there are tenurial disincentives resulting in lower input and output per acre among the tenanted holdings and if the proportion of area leased in increases with size, then the decline in output per acre with size could be explained partly by the operation of tenurial disincentives. The need to test this hypothesis was suggested by Khusro.[11]

But in areas where Farm Management Studies are conducted, the proportion of land taken on lease declines as the farm size increases.[12] With a view to finding out the effect of tenancy on output when size of holding and the proportion of area irrigated are held constant, the equations given below were derived for data relating to Bombay:

$$\log Y = 1.55 + \underset{(0.09)}{0.88} \log X_1 + \underset{(0.14)}{0.80} \log I - \underset{(0.10)}{0.02} \log T \quad \text{(Ahmednagar)}$$

$R^2 = 0.61$; N = 80

$$\log Y = 1.96 + \underset{(0.08)}{0.65} \log X_1 + \underset{(0.12)}{0.80} \log I - \underset{(0.07)}{0.01} \log T \quad \text{(Nasik)}$$

$R^2 = 0.55$; N = 80

Where Y = farm output in rupees; X_1= farm area in acres; I = proportion of holding irrigated; and T= proportion of holding leased in. The following dummies are assigned to represent the per cent of area leased in: Upto 25 per cent of cultivated area = 1; upto 50 per cent = 2; upto 75 per cent = 3; and above 75 per cent = 4. Although the coefficients for T have negative signs, they are not statistically significant. This looks theoretically plausible in the context of Sen's cheap-family-labour-based explanation. Although permanent improvements will be discouraged on the tenanted holdings, current inputs (particularly family labour) may be applied liberally, because so long as the marginal net product is positive for such 'free' resources, their application will add to the income of the tenant even when the rent bears a fixed proportion to the gross produce. According to the Farm Management Studies, the ratio of wage costs to output is the lowest for tenanted holdings,[13] as they generally belong to small and middle size-groups. The effect of tenancy on the behaviour of productivity with size cannot, therefore, be considered very significant.

Management-Based Explanation

The labour-based explanation with the necessary modification may hold true for a certain range in farm size where the necessity of hiring labour in the slack season does not arise. But as Khusro pointed out, the family labour of the farmer finds full employment on a size ranging from 7.5 acre to 10 acre in different regions of the country and the need to employ permanent farm servants arises beyond this point.[14] So long as some permanent labour is hired to work in the slack season also, the employment of the marginal unit of labour will be within a profitable range even if a part of labour is supplied by the farm family and the cheap-labour argument in the sense discussed earlier ceases to be relevant. Dipak Mazumdar's proposition, in principle, is the same but he has in mind the small family farms who hire casual labour for peak-season operations, when the marginal supply price of family labour is not lower than the ruling wage rate.[15] He concedes that among small farms, family labour may be applied liberally in the slack period as their marginal supply price is less than the wage rate. His argument, in its essence, lends support to Sen's labour-based explanation. It is evident from the findings of the Farm Management Studies that inputs and output per acre decline consistently with size for holdings above 10 acre. Some more evidence on this is provided by the following functions derived for farms of: (i) West Godavari, operated wholly with hired labour, both casual and permanent; and (ii) Karimnagar, where some permanent labour is necessarily hired:

(i) $\log Y = 2.93 + \underset{(0.01)}{0.90} \log X_1$ (Andhra: West Godavari)

$R^2 = 0.94; \quad N = 28$

(ii) $\log Y = 2.47 + \underset{(0.04)}{0.74} \log X_1$ (Andhra: Karimngar)

$R^2 = 0.94; \quad N = 20$

The coefficients are significant at 1 per cent level and are significantly different from unity. That the decline in output per acre with size becomes less marked after correcting for irrigation among these farms also, is indicated by the following equations:-

(i) $\log Y = 1.29 + \underset{(0.03)}{0.94} \log X_1 + \underset{(0.20)}{0.83} \log I$ (West Godavari)

$R^2 = 0.98; \quad N = 28$

(ii) $\log Y = 2.19 + \underset{(0.08)}{0.79} \log X_1 + \underset{(0.17)}{0.14} \log I$ (Karimnagar)

$R^2 = 0.95; \quad N = 20$

The inclusion of irrigation improved the R^2. The coefficients for land are raised significantly, indicating that the decline in output per acre with size is explained to a certain extent by the declining significance of irrigation. The variation of labour and capital inputs (X_2) in relation to size (X_1) is given by the following equations:

(i) $\text{Log } X_2 = 2.58 + \underset{(0.01)}{0.85} \log X_1$ (West Godavari)

$R^2 = 0.95; \quad N = 28$

(ii) $\text{Log } X_2 = 2.26 + 0.81 \log X_1$ (Karimnagar)
$\qquad\qquad\qquad\qquad (0.01)$

$R^2 = 0.89; \quad N = 20$

The coefficients are statistically significant and are significantly different from unity, indicating that the labour and capital inputs vary less than proportionately to size even among farms where permanent labour is partly or wholly hired.

It has been observed above that the labour-based explanation, in the sense of its application beyond the profitable range (and not in the formal sense of a decline in its input per acre with size) is not relevant for farms operated partly or wholly with hired permanent labour. It has also been argued that fertility and tenancy factors are not significant in explaining the output behaviour with size. If one agrees with Sen on what seems to be a reasonable proposition, *viz.*, that the availability of capital cannot be a bottleneck for larger farms,[16] then the productivity behaviour can be explained only in terms of the management factor and the income-leisure preferences among the larger farms. Sen is prepared to concede only the qualitative aspect of managerial factor reflected in the 'loving care' that the small property inspires.[17] That he regards management as one of the independent variables along with capital and labour is evident from his treatment of the management or technique-based explanation as different from the labour-based and capital-based explanations.[18] Such a formulation ignores that capital and labour inputs per acre are themselves dependent on the management factor. Since farming processes require close personal attention at each stage and since the scope for division of functions and the delegation of responsibility is very limited, managerial and supervisory bottlenecks arise at comparatively lower levels of farm size. This stage will be reached much earlier under labour-intensive techniques of cultivation, as the human problems of management can be expected to be particularly acute in such a situation.[19] Sen thinks that such problems can be overcome by splitting the holdings and cultivating them in small bits.[20] In fact most Indian farms are cultivated in small bits and very often the difficulties of supervision bear a direct relationship to the number of such bits. There is no doubt that as the farm size increases, say, from 20 acre — which can be regarded as a convenient managerial unit under Indian conditions — to 100 acre or five times, the management input cannot be increased in the same proportion because a hierarchy of paid supervisors over paid labour is not a profitable proposition. Managerial bottlenecks could thus affect productivity with size via the decline in the labour and capital inputs per acre. This is understandable because land is more or less institutionally fixed for big farmers and does not involve payment of rent for its use. It is also possible that income itself is an important variable motivating the small farmers to work the land intensively and introducing an element of indifference among big farmers, especially when such farmers are accustomed to the traditional consumption patterns and show greater preference for leisure and non-economic pursuits in the villages. Also, in view of the very high rates of interest prevailing in the villages, big farmers may find direct money-lending operations more convenient than the intensive cultivation of land.

Implications

Sen concludes from his analysis that difference in the size of holding *as such* is not very important and what makes the crucial difference is whether farming is family-based or wage-based.[21] The above analysis suggests, on the other hand, that the size of holding *as such* is an important factor affecting productivity. Labour-based as well as the management-based explanations would imply that measures to effect the transfer of land from the big to the small and medium farms through sales and through tenancy on a commercial basis are desirable. The

Management-based explanation would imply further that, if a substantial reduction of large holdings through such measures is ruled out, then the encouragement of mechanised processes would be necessary as this would help to intensify the inputs among large farms by overcoming managerial difficulties.

Notes

1. Directorate of Economics and Statistics, Ministry of Food and Agriculture, Government of India, *Studies in Economics of Farm Management*. The studies conducted so far relate to seven regions *viz.*, Uttar Pradesh, Punjab, West Bengal, Madras, Bombay, Madhya Pradesh and Andhra Pradesh.
2. See Sen, A.K., "An Aspect of Indian Agriculture," *The Economic Weekly*, Annual Number, February, 1962; Mazumdar, Dipak, "On the Economics of Relative Efficiency of Small Farmers," *The Economic Weekly*, Special Number, July, 1963; Sen, A.K., "Size of Holdings and Productivity," *The Economic Weekly*, Annual Number, February, 1964; Khusro, A.M., "Returns to Scale in Indian Agriculture," *The Indian Journal of Agricultural Economics*, October-December, 1964; and Hanumantha Rao, C.H., "Farm Size and the Economies of Scale," *The Economic Weekly*, December, 14, 1963. Implicit in all these analyses is the recognition of the utility of measuring productivity in terms of a single factor like land because of the scarcity of its supply under Indian conditions.
3. Sen, "Size of Holdings and Productivity," *op. cit.*
4. Source of data for Bombay: *Studies in Economics of Farm Management in Bombay*, 1955-56, *op.cit.* Individual farm data in the published form are available for the Bombay study only. Source of data, for Andhra Pradesh: Survey conducted by the author in connection with a study undertaken at the Institute.
5. Sen, "An Aspect of Indian Agriculture," *op. cit.*
6. *Ibid.*
7. The costs on bullocks and implements as a proportion to the value of gross output are found to decline with the increase in farm size in all the regions studied. See, Hanumantha Rao, C.H., *Agricultural Production Functions, Costs and Returns in India*, Asia Publishing House.
8. Sen, "Size of Holdings and Productivity," *op. cit.*
9. Khusro, "Returns to Scale in Indian Agriculture," *op. cit.*
10. Lakdawala, D.T., *Justice in Taxation in India*, 1946.
11. Khusro, "Returns to Scale in Indian Agriculture," *op. cit.*
12. See, Hanumantha Rao C.H., *Agricultural Production Functions, Costs and Returns in India, op. cit.* Khusro's findings regarding the positive association between the proportion of area leased in and the size of holdings is based on sources other than Farm Management Studies, See his "An Analysis of Agricultural Land in India by Size of Holding and Tenure (mimeographed)."
13. Hanumantha Rao C.H., *Agricultural Production Functions, Costs and Returns in India, op. cit.*

14. Khusro, "An Analysis of Agricultural Land in India by Size of Holding and Tenure," *op. cit.*
15. Mazumdar, "On the Economics of Relative Efficiency of Small Farmers," *op. cit.*
16. Sen, "Size of Holdings and Productivity," *op. cit.*
17. *Ibid.*
18. *Ibid.*
19. For a detailed discussion of this aspect, see, Hanumantha Rao C.H., "Farm Size and the Economies of Scale," *op. cit.*
20. Sen, "Size of Holdings and Productivity," *op. cit.*
21. Sen, "An Aspect of Indian Agriculture," *op. cit.*

This paper has been prepared at the Institute of Economic Growth, Delhi, and published in *Indian Economic Review*, October, 1966. The author is grateful to Miss Har Bhajan Kanwar and Mr. Pushpendar Kumar who did the statistical work for this paper. The author is indebted to Dr. Dharam Narain, Dr. A.M. Khusro, Dr. K. Krishnamurty and Dr. K.A. Naqvi for their very useful comments on an earlier draft of this paper.

20

Uncertainty, Entrepreneurship and Sharecropping in India

The Problem

Economists have, by and large, viewed sharecropping arrangements as inefficient and owner farming and leasing on a fixed-rent basis as conducive to the efficient use of land.[1] They tend to regard the tenurial arrangements as exogenous constraints determined largely by custom and tradition and do not attempt to analyse how and to what extent the prevailing lease arrangements might themselves have been influenced and shaped by economic factors.[2] Such an understanding is especially necessary for economies where, despite the freedom to choose between different contractual arrangements, owner farming through hired labour as well as different types of lease arrangements are found to coexist and persist through time.[3] The sharecropping system represents a deviation from the widely prevalent system where the individual entrepreneurs — whether in industry or in agriculture — guarantee fixed rates of payment to factors in advance of production. The system of remunerating factors according to predetermined proportions of output needs, therefore, to be explained. This paper attempts to explain the coexistence of different types of lease arrangements in terms of the varying significance of the entrepreneurial functions with respect to different crops, which are themselves influenced by the degree of economic uncertainty faced by farmers.

The Hypothesis

Farm lease contracts may be defined as institutional devices for allocating risk among landlords and tenants (Schultz 1968).[4] Cash rents guaranteed in advance of production imply that the risks of production are shared entirely by the tenants, while crop-sharing rentals indicate the distribution of such risks among the tenants and the landlords in proportion to their respective shares in output. Fixed-kind rents settled in advance of production imply the sharing of price uncertainties but allocate the yield risks entirely to the tenants.

Under conditions of perfect knowledge when the economic system is in equilibrium, there is no particular economic basis for classifying the gross receipts of the firm into those pertaining to the hired and the hiring factors or the contractual and residual returns. The basis for this distinction is provided by the existence of uncertainty or incomplete knowledge regarding the actual realisable input-output rates as well as the prices and the consequent necessity of acting upon judgment or opinion rather than knowledge (Knight 1965, pp. 264-90). The guaranteeing of fixed contractual returns and the emergence of profit as the residual are, therefore, necessary consequences of uncertainty involving entrepreneurial judgment and decision making.

But situations of uncertainty may exist without much scope for entrepreneurship. It is possible, for instance, that the returns from any two competing crops reveal a high degree of variance in the past and yet the returns from one of them are always higher than from the other. Insofar as past experience is a guide to decision making, this represents a situation where the distributions of the expected returns from the crops concerned do not overlap. In such a situation, there is no effective uncertainty facing the decision maker, so that the choice of the product may be made simply on the basis of maximum profit (Alchian 1950). Yet, uncertainty may exist in the sense that the returns from the crop chosen can be identified only with a probability distribution of potential outcomes and not with a unique outcome (Alchian 1950). On the other hand, if the distributions of expected returns from alternative crops overlap, then uncertainty may be said to exist for decision making. Therefore, the existence of entrepreneurial function is indicated not by the existence of uncertainty as such but by the scope for decision making in the face of uncertainty.

The existence of a significant scope for entrepreneurship may occasion fixed-cash rents, because the latter at the same time permit the tenants to capture the returns expected in consequence of their decision making and protect the suppliers of land services against the possible risks arising from the production decisions of the tenants.

Under uncertainty situations where the scope for entrepreneurial functions is limited, there may be an incentive for the tenants to reduce the variance of their incomes by shifting part of the uncertainty to the suppliers of land services through sharecropping arrangements.[5]

Since conditions of production in agriculture are extremely varying and important decisions must be made at different intervals, large farms may face managerial (supervisory) diseconomies (Robinson 1962, p. 44). Also, since indivisibilities are relatively few in agriculture, technical gains on a large scale may be negligible (Schultz 1965, p. 122), especially under labour-intensive farming. Therefore, in situations where the element of uncertainty is smaller and entrepreneurial functions are relatively unimportant and where, therefore, the farm takes on the character of a managerial or supervisory unit, the landlords may find it profitable to lease out (on a share-rent basis),[6] beyond a point, the land they own rather than cultivate the entire holding through hired labour.[7] The incentive to lease out on a share basis, however, would exist only if the landlords can expect that the share rents, on the average, would not be lower than the fixed rents or the net income from own cultivation with hired labour.[8]

Marshall, whose ideas on the inefficiency of sharecropping are popular among economists, stated that the rent under this system would be lower than the fixed rent or net income from own cultivation if the tenants were free to restrict their inputs.[9] He, nevertheless, thought that the landlord ensured the desired allocation of labour from share tenants by spending "much time and trouble either of his own or of a paid agent, in keeping the tenant to his work" (Marshall 1956). However, given the managerial limitations which may induce the landlords to lease out land and the free institutional setting, both of which limit the capacity of the landlords to supervise the tenants' allocation of inputs, the incentive to lease out would exist if the production functions for share-rented crops are such as not to leave much scope for share tenants to restrict the use of inputs.[10]

Cheung's theory of the choice of contracts (Cheung 1969) refers to the "gains from risk dispersion" but omits any consideration of the scope for decision making in the face of uncertainty and the nature of the production functions. His theory would thus suggest that share contracts can be expected to be more widespread in areas characterised by a high degree of uncertainty than in areas of relative certainty, if transaction costs are equal (Cheung 1969, p. 70). However, as we have argued above and tested against the Indian data in the next section, relative economic certainty in the sense of limited scope for decision making seems to be necessary for the prevalence of sharecropping. Otherwise, individual anticipations regarding input-output rates and prices may differ, making it difficult for the parties concerned to arrive at an agreed choice as to the product mix and the amount of inputs to be committed. Our results suggest further that situations of high uncertainty may necessitate fixed-cash rents.

Cheung notes "the tenant's incentive to use an amount of input less than that stipulated in a share contract" (Cheung 1969, p. 67) and suggests the role of transaction costs (which include the costs of "enforcing" the stipulated input) in determining the lease arrangements (Cheung 1969, pp. 67-68). Thus, his reasoning in this respect, like that of Marshall, rests on the implicit assumption that the proportion between land and labour inputs can be varied over a significant range for share-rented crops. However, in the case of production functions characterised by relatively inflexible input combinations, the costs of enforcing tenant's input would be lower and the incentive for share contracts would be greater. This condition would seem necessary for areas where sharecropping is extensive and where landlords prefer to lease out land, owing to managerial or supervisory diseconomies of own cultivation with hired labour.

The Evidence

A. Incentives for Enterprise or Product Substitution

Sharecropping is extensive for the rice crop in the eastern and southern regions of India and for wheat in the northern regions. A recent study (Ladejinsky 1965) relating to the districts where the Intensive Programme of Agricultural Development (Package Programme) is in progress revealed that the sharecropping system of tenure is extensive for the rice crop in West Godavari (Andhra) and Thanjavur (Tamil Nadu) in the south and for the wheat crop in Ludhiana (Punjab) in the north. In all these places, 50 per cent of the gross output is paid as rent (Ladejinsky 1965). The available evidence suggests that the scope for allocative decisions is very much limited in the rice and wheat tracts.

In a careful study of the response of rice areas in relation to the relative price of rice lagged one year for several rice-growing regions of undivided India, for example, Bengal, Bihar, Orissa and Madras for the period 1900-1939, Narain (1965) found that with the exception of aus rice (autumn crop), which competes with jute in Bengal,[11] the areas under rice cultivation responded little to changes in its relative price.[12] Narain concluded that "conditions which render a soil suitable for rearing rice often render it unsuitable for bearing an alternative crop. And, to the extent that the areas under rice are in this sense specific to it, the allocative function of price breaks down" (Narain 1965, p. 109). The "specificity" of rice fields seems to be a rather widespread phenomenon. Smith (1937, p. 160) observed that a rice field "is unfit either for corn, or pasture, or vineyard, or indeed for any other vegetable produce that is very useful to men; and the lands which are fit for those purposes are not fit for rice."[13]

All this evidence need not lead to the inference that fields suited to rice cannot be made to grow other crops either in the short run or through drainage in the long run. Nor should it imply that lands growing other crops cannot be turned to rice in the long run through irrigation. Nevertheless, the inelasticity of a rice area with respect to its relative price may suggest that, given the prevailing prices and yields of competing crops and the variations in them, rice has a decided long-run relative advantage, so that there is no economic incentive for the farmer to shift an area from rice to alternative crops.

In a similar investigation for wheat in Punjab, United Provinces, Central Provinces and Berar, and Bombay and Sind — the four leading provinces of wheat cultivation, accounting for approximately nine-tenths of the country's wheat area — Narain found scarcely any positive relationship between area and the deflated price of wheat, except in the Meerut Division of the United Provinces, where there was evidence of area shifts between wheat and sugarcane (Narain 1965, pp. 135-57). In his study for Punjab extending from 1914 to 1944, Krishna (1963) shows that, whereas the acreage elasticities of cotton and maize were significantly higher than the comparable estimates by Nerlove (1958) for the United States, the short-run acreage elasticity of wheat with respect to its expected relative price

(0.08) was the lowest among the acreage elasticities estimated for the various crops and was significantly lower than Nerlove's estimate of 0.48 for wheat in the United States for the period 1901-32 (Nerlove 1958, p. 202). The insensitivity of a wheat area to its relative price is understandable, because in the *rabi* season (November-March), when wheat is grown, there is no effective competition from cash crops. Gram is the other principal crop grown in this season, and its area movements do not bear an inverse relationship to those of wheat, owing to the sympathy between their prices and also because climatic factors affect them in the same direction (Narain 1965, pp. 135, 158).

B. Incentives for Factor Substitution

Under relative economic certainty, the *ex ante* returns would not diverge significantly from the *ex post* returns. On the other hand, in situations characterised by uncertainty and a significant scope for entrepreneurial functions, one should expect a significant divergence between the expected and the realised returns. Owing to the interpersonal differences in judgments based on differing anticipations regarding prices, yields, *etc*., the variation in profit between individuals can be expected to be larger in this situation. An attempt has, therefore, been made to study the variance in profit earned by farmers operating under different lease contracts. The West Godavari district of Andhra Pradesh State provides an interesting case for a study of this kind, because within the same district, share-lease and cash-lease arrangements coexist, the latter being negligible in the rice zone and predominant in the tobacco zone (Table 20.1). Also, the rice crop, for which the share-lease system is extensive, is a major marketed or cash crop of the region, so that the share-lease system cannot readily be explained in terms of the subsistence nature of the crop. Detailed data on farm management are available for this district in the "Studies in Economics of Farm Management" (Government of India 1963). These studies were carried out for three years, 1957-58, 1958-

Table 20.1

Farm Lease Contracts in West Godavari, India

	Percentage of Tenanted Area Under			
Zone	*Share Leases*	*Cash Leases*	*Fixed-Kind Rent*	*Total*
Rice	42	2	56	100
Tobacco	14	56	30	100

Source: Government of India (1963, table 2.7a).

59 and 1959-60, in seven villages of the rice zone and three villages of the tobacco zone. In all, 104 farms (sixty-seven in the rice zone and thirty-seven in the tobacco zone) were randomly selected each year and were studied through the cost-accounting method.[14] Net profits were estimated in these studies by deducting from gross returns all the expenses actually incurred as well as the imputed values of the services of own resources, for example, land, labour and farm equipment. Profit so defined[15] approximates to the difference between the *ex ante* and *ex post* earnings. In these studies, profits of individual farms in each zone were aggregated and averaged across the villages according to eight classes of farm size for each year separately. We have arranged these figures according to seven size classes of profit per acre in order to study their frequency distributions (Table 20.2).

The variance of profit is much lower for rice compared with tobacco. In the case of rice, as much as 50 per cent of the observations on profit (or loss) per acre lie within the range of ± Rs. 25 and about 80 per cent within the range ± Rs. 50. The corresponding percentages for tobacco are 14 and 23.

The data on irrigation, pattern of irrigation, variation in rainfall, and prices suggest that the cultivation of tobacco is done under a high degree of uncertainty, whereas rice farmers operate under conditions of relative economic certainty[16] with a relatively smaller farm (Table 20.3).

Table 20.2

Frequency Distribution of Profit Per Acre, West Godavari (Combined for the Years 1957-58, 1958-59 and 1959-60)

	Rice		Tobacco	
Size of Profit (Rs. Per Acre)	*Frequency*	*%*	*Frequency*	*%*
– 100	...	...	6	28.5
– 100 to – 50	4	16.6	...	...
– 50 to – 25	5	21.0	1	4.8
– 25 to + 25	12	50.0	3	14.3
+ 25 to + 50	2	8.2	1	4.8
+ 50 to + 100	1	4.2	3	14.3
+ 100	...	...	7	33.3
Total	24	100.0	21	100.0

Crop	*Mean*	*Standard Deviation*	*Coefficient of Variation*
Rice	-13.0	39.0	300.0
Tobacco	21.0	153.0	729.0

Source: Government of India (1963, tables 5.10, 5.23).

The crop figures of inputs, output, and profit (as defined above) per acre relating to large owner-operated farms which leased out part of their land, as well as the figures of rents received by them, taken from a study made by Rao (1966a) in the same district in 1958-59, are given in Table 20.4. Rice lands were leased out on a share basis, whereas lands growing sugarcane and chillies were leased out on a cash rent basis. Profit per acre is much higher for sugarcane and chillies than for rice. Rents received constitute a much smaller proportion of gross output in the case of these crops compared with rice. These facts may suggest the significance of the entrepreneurial factor — the scope for decision making as well as profitability — from the point of view of tenurial arrangements. Cash rents for more profitable crops are understandable because the tenant can capture the returns himself on the investment made.

Table 20.3

Irrigation, Rainfall, Prices, and Average Size of Holding in West Godavari

Variable	*Rice Zone*	*Tobacco Zone*
Cultivated area irrigated (%)	77.5	36.0
Distribution of irrigated area (%, by source):		
Canals	91.5	50.3
Tanks	5.7	11.9
Wells	2.8	37.8
Average annual rainfall (inches)	48.0	41.6
Coefficient of variation in rainfall (1952-53 to 1959-60)	13.3	22.1
Coefficient of variation in wholesale prices (1949-58)	18.9	34.9
Average size of holding (acres)	6.7	11.2

Source: Government of India (1963).

Table 20.4

Relative Profitability of Crops in West Godavari, India (1958-59)

	Value per Acre (Rs.)				
	Gross Output	*Cost of Production*	*Value*	*Rental Value*	*% of Rental Value to Gross Output*
Rice (first crop)	471	432	39	274	58.2
Rice (second crop)	419	418	1	214	51.1
Chillies	1,041	815	226	360	34.5
Sugarcane	2,128	1,282	846	500	23.1

Source: Rao (1966a, pp. 86, 88).

In the rice zone, tenants belong mainly to the small and middle-size holding groups, and those who lease out belong to the large holding group. In the tobacco zone, on the other hand, those who lease out belong mainly to small and middle-size holding groups,[17] and in the case of the large holding group, area leased in constitutes a substantial proportion of the holding (Table 20.5). Thus, the average size farm is significantly larger in the tobacco zone (Table 20.3) compared with the rice zone. Since large farmers in the rice zone as well as small farmers in the tobacco zone are free to lease-in land, these facts may suggest the incentive on the part of the large landowners in the rice zone to benefit from managerial economies which, under relative economic certainty, may outweigh the gains of own cultivation with hired labour. Thus, it appears that rice is sharecropped by small farmers essentially for augmenting factor incomes through the fuller use of own resources, such as family labour and bullocks, whereas large farmers in the tobacco zone lease in area at fixed-cash rents with a view to earning profit.[18]

Table 20.5

Percentage of Operated Area Leased-in and Leased-out According to Size Group of Holding, West Godavari (Combined for the Years 1957-58, 1958-59, and 1959-60)

	Rice Zone		*Tobacco Zone*	
Size of Operated Holding (Acres)	*Leased-In*	*Leased-Out*	*Leased-In*	*Leased-Out*
Up to 5	28.6	2.3	14.8	6.2
5 to 20	40.3	3.5	41.3	7.3
Above 20	2.4	21.9	20.8	1.2
Overall	23.8	9.9	31.2	6.1

Source: Government of India (1963, tables 3.1, 3.5).

C. Resource-Use Efficiency

An attempt is made below to study the comparative efficiency with which land is cultivated among owner-operated and share-rented farms in the rice zone during 1957-58 and 1958-59.[19] All those who lease-in area — part as well as pure tenants — are classified as tenants.[20] There is a high intercorrelation among all the inputs — land, labour and capital (Table 20.6).[21] Land and labour inputs show the highest intercorrelation between both the tenurial groups of farms, and this correlation does not differ between the two groups. These results suggest that factor proportions, especially between land and labour, are relatively inflexible irrespective of the tenure systems. This result, though interesting, should not be surprising for monoculture of deltaic rice regions. The possibility of varying labour input may exist for certain operations, especially for weeding and inter-culturing. But, in view of their complementarity with operations which precede and follow them involving inflexible combinations, it is likely that there is not much economic incentive for share tenants to restrict inputs for such operations.[22]

Table 20.6

Correlations Between Input Variables for Rice, West Godavari

Year	Type of Tenure	Number of Observations	Correlation Between		
			X_1, X_2	X_1, X_3	X_2, X_3
1957-58	Owner-operated	37	0.92	0.83	0.91
1957-58	Share-rented	30	0.92	0.89	0.94
1958-59	Owner-Operated	33	0.94	0.76	0.87
1958-59	Share-rented	37	0.91	0.88	0.91

Note: X_1 = land input; X_2 = labour input; X_3 = capital input.

An attempt to fit the Cobb-Douglas type of production function to these data did not yield reliable estimates of the elasticity coefficients of individual inputs, owing to the very high inter-correlation among the input variables. It was, therefore, decided to estimate the production functions with land input alone as the independent variable. The function used is $Y = ax^b w$, where Y is output, x is land input, and w is the random error. The coefficients were estimated through the method of least squares. Variation in land input explains around 90 per cent of the variation in output (Table 20.7). These coefficients for land should be interpreted as reflecting the joint effect of all other inputs which may be varying along with land input. Output per acre at the margin, that is, for a unit increase in land at its geometric mean,[23] is higher among the owner-operated farms compared with share-rented farms, indicating that, on the average, land is cultivated more intensively by owner-operated than by share-rented farms. However, the elasticity coefficients indicate diminishing marginal productivity of land among owner-operated farms as against constant returns under sharecropping. Since as much as 80 per cent of the land leased-out belongs to farms larger than 20 acre and since small and middle-size farmers (5 to 20 acres) account for three-fourths of the land leased-in (Table 20.8), it would be reasonable to suppose that these two groups are the relevant tenurial alternatives for the comparison of allocation efficiency. The earlier studies on allocation efficiency under owner farming *vis-a-vis* sharecropping treated all owner-operated farms as a single category and came up with the conclusion that land is, on the average, cultivated more intensively among owner-operated than among share-rented farms, which was regarded as sufficient evidence in support of Marshall's proposition concerning resource-use efficiency under sharecropping.[24] But the

Table 20.7

Production Functions for Rice, West Godavari

Year	*Type of Tenure*	*Number of Observations*	*Constant*	*Regression Coefficient of X (Land)*	R^2	*Output per Standard Acre at Margin (Rs.)*
1957-58	Owner-operated	37	1.85 (0.30)	0.84* (0.05)	.90	210.5
1957-58	Share-rented	30	1.39 (0.40)	0.91 (0.06)	.90	202.7
1958-59	Owner-operated	33	1.50 (0.29)	0.91** (0.05)	.93	231.9
1958-59	Share-rented	37	0.50 (0.41)	1.03 (0.06)	.88	203.6

Note: All regression coefficients are statistically significant at P=.01. Figures in parentheses are standard errors of the respective coefficients.

* Significantly different from unity at P = .05.

** Significantly different from unity at P = .10.

above conclusion would be significantly modified if we note the widely prevalent phenomenon of diminishing returns to land among large owner-operated farms in India (Rao 1966b) and compare the allocation efficiency of groups leasing-out and those leasing-in land.

Table 20.8

Percentage Distribution of Operated Area, Area Leased in, and Area Leased-Out According to Size of Operational Holding, Rice Zone, West Godavari (Combined for the Years 1957-58, 1958-59 and 1959-60)

Size of Holding (Acres)	*Operated Area*	*Area Leased-In*	*Area Leased-Out*
Up to 5	16.8	20.0	4.0
5 to 20	46.9	76.3	16.0
Above 20	36.3	3.7	80.0

Source: Government of India (1963, tables 3.1, 3.5)

Marginal products of land have, therefore, been estimated at each holding size level for owner farms as well as share-rented farms (see Figure 20.1). There is a decline in the marginal productivity of land with an increase in the size of holding among owner-operated farms.[25] Over a wide range, the marginal productivity of land among the share-rented farms is higher than owner-operated farms of corresponding size and is not lower than large owner-operated farms, which account for the bulk of land leased-out. A separate study made by Rao (1966a) in the same district enables a comparison of the productivity of share-rented plots and large owner-cultivated farms; Table 20.4 shows that rents per acre for rice received from sharecroppers constituted 51 to 58 per cent of gross output per acre realised on the large owner-cultivated farms. Since share rents constituted 50 per cent of gross output of tenanted plots, these data suggest that sharecroppers cultivated the land more intensively than the large owner-operators.[26]

Figure 20.1

Marginal Product Curves of Land for the Rice Zone, West Godavari, India

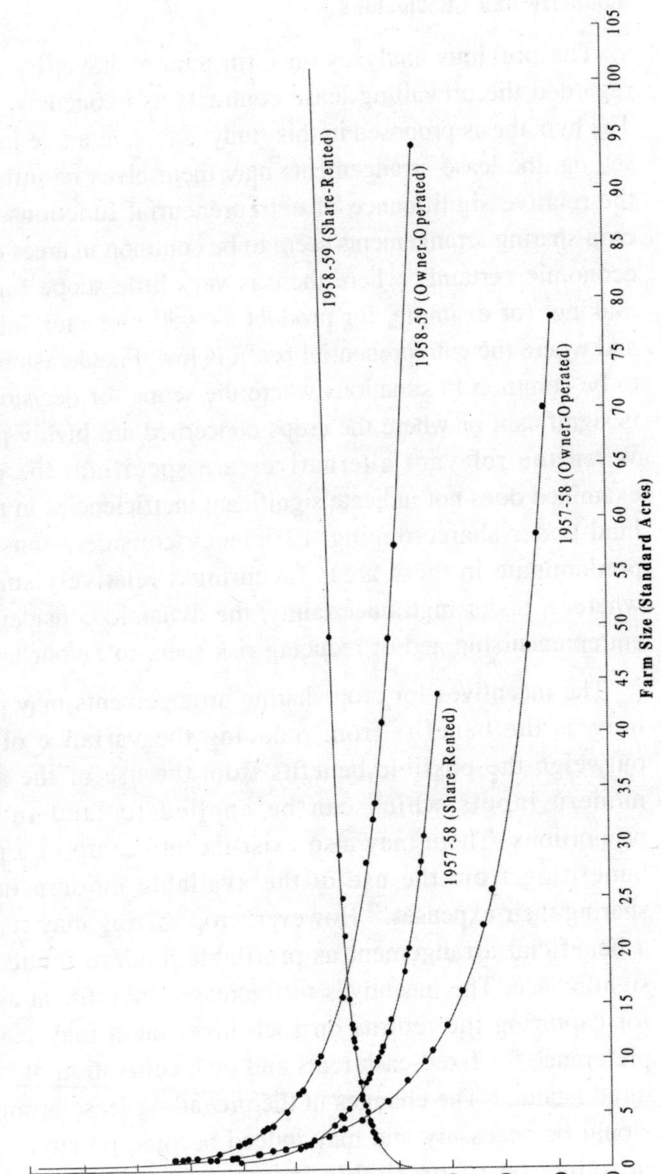

Summary and Conclusions

The previous analyses on farm tenures have, by and large, regarded the prevailing lease contracts as exogenous variables. The hypothesis proposed in this study is that, in a free institutional setting, the lease arrangements may themselves be influenced by the relative significance of entrepreneurial functions. In India, crop-sharing arrangements seem to be common in areas of relative economic certainty where there is very little scope for decision making, for example, for product as well as factor substitution, and where the entrepreneurial profit is low. Fixed-cash rents seem to be common in situations where the scope for decision making is significant or where the crops concerned are highly profitable. When the relevant alternatives are specified, the evidence examined does not indicate significant inefficiencies in the use of land under sharecropping. Efficiency considerations seem to predominate in these areas favouring a relatively small farm, whereas, under high uncertainty, the dynamic considerations of entrepreneurship and of reducing risk seem to favour larger size.

The incentives for crop-sharing arrangements may persist as long as the benefits from reducing the variance of income outweigh the possible benefits from the use of the available modern inputs which can be applied to land in varying proportions. There may also exist incentives up to a point for benefiting from the use of the available modern inputs by sharing their expenses.[27] However, crop sharing may cease to be a beneficial arrangement as profitable modern inputs assume significance. The incentives for increased investment as well as for capturing the returns on such investment may lead to the preference for fixed-cash rents and own cultivation of land with hired labour.[28] The changes in the prevailing lease arrangements would be necessary and may indeed become possible when the introduction of profitable inputs raises the incentives for investment and widens the scope for decision making.

Notes

This study was undertaken and pursued at the Faculty of Economics, University of Chicago, with support from the Ford Foundation, during 1966 and 1967. It was pursued further at the Institute of Economic Growth, Delhi. I am grateful to T.W. Schultz, D. Gale Johnson, Dharm Narain, and P.K. Bardhan for their helpful criticisms and suggestions. I am also grateful to S.V. Sethuraman, Ajit Singh Vats, and Pushpendra Kumar for their generous help and cooperation. I also benefited from the comments of Jagdish N. Bhagwati and Sukhamoy Chakravarty on an earlier draft of this paper (Bhagwati and Chakravarty 1969). However, I am alone responsible for the deficiencies of the paper.

1. For a comprehensive discussion of the views of many classical and neoclassical economists on this subject, see Johnson (1950).

2. A significant exception is the pioneering attempt by Schultz (1940) to explain the preference of farmers in the United States to lease in rather than buy and own land in terms of capital rationing — which itself is rooted in economic uncertainty — practiced in various degrees by creditors as well as farmers. Cheung's work (1969, p. 71) represents another major attempt to explain the prevailing tenurial arrangements in terms of economic factors: "The choice of contracts is determined by weighing the gains from risk dispersion and the transaction costs associated with different contracts."

3. In India, for instance, despite public legislation to discourage share tenancy, many resident landlords in rice-growing as well as wheat-growing regions continue to lease out land when they are in fact encouraged to cultivate the land themselves through hired labour. In respect of these crops, they enter into "off the record" share arrangements with the tenants when the system of cash renting is in vogue for certain other crops in the same regions (see Ladejinsky 1965). An interesting example of the coexistence of share and cash rents is provided in the corn belt of the United States; in areas where "crop-rotations have become diversified and livestock enterprises have expanded, it has become customary to pay a cash rent for land in hay and pasture in addition to the crop-share rent" (Shickele 1941).

4. Schultz emphasises in this paper the functional character of institutions and the need to explain the prevalence of particular institutions in terms of economic factors instead of treating them as exogenous variables. He suggests further that institutions may respond, though with varying lags, to the dynamics of economic growth.

5. Shifting of uncertainty may not be without a price. Since stable income flows are generally preferred to the fluctuating ones of the same normal or average value, the normal value of the going share rents may be higher than would be the case if share rents were converted into fixed ones.

6. That landlords too may prefer share rents to fixed rents in such situations appears plausible because the normal or average value of share rents could be higher than fixed rents for reasons mentioned earlier and in view of the greater capacity of landlords for uncertainty bearing arising from their better wealth position.

7. Under high degree of uncertainty, however, managerial diseconomies of large size may be outweighed by the benefits from the reduction of risk due to the operation of the law of large numbers (Arrow 1951).

8. Share rents acceptable to landowners could be lower than the net income from own cultivation depending upon the subjective value attached to the stable income flows.

9. "When the cultivator has to give to his landlord half of the returns to each dose of capital and labour that he applied to the land, it will not be to his interest to apply any doses the total return to which is less than twice enough to reward him. If, then, he is free to cultivate as he chooses... he will apply only so much capital and labour as will give him returns more than twice enough to repay himself: so that his landlord will get a smaller share even of those returns than he would have on the plan of a fixed payment" (Marshall 1956, p. 644). The diagram below illustrates this proposition with labour as a variable resource applied to a unit of land. Under own cultivation, $\partial y/\partial x=w$, output from a unit of land, say an acre, is equal to the area *oxw'y*, and net income is equal to *ww'y*, which would also be the amount under a fixed-rent contract. Under sharecropping where rent constitutes 50 per cent of output, it will not be to the advantage of the tenant to extend the input beyond *x'* where $\partial y/\partial x=2w$, output per acre will not exceed the area *ox'y'y*, and rent will not exceed one-half of this area, or *pmy'y*, which is less than *ww'y*. The share tenant can get additional income to the extent of the area *wmp* as compared with what he would otherwise get as a wage earner.

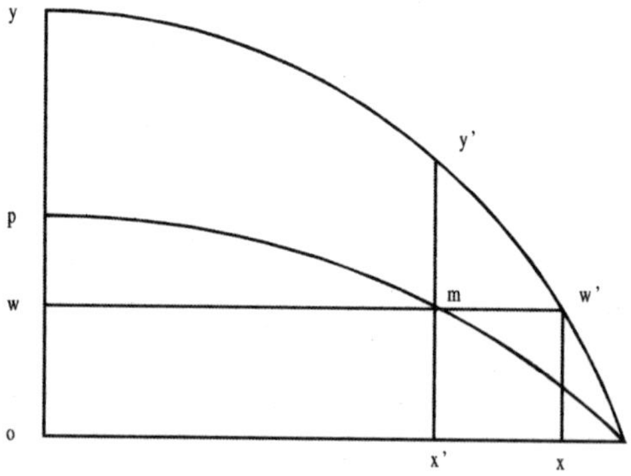

10. Johnson (1950) describes, with reference to the Great Plains of the United States, the circumstances under which the landlord has a "reasonable assurance that the land will be farmed with appropriate intensity."

11. Venkataraman (1958) estimated an elasticity of 0.46 for jute area with respect to its relative price, the main competing crop being rice. It is interesting to note that cash leases were found to be more numerous in areas where the cultivation of jute was extensive than in areas where rice cultivation was extensive (Government of India 1957, p. 8).

12. In his investigation of the acreage response of rice to its relative price in East Bengal (East Pakistan) for the period 1948-61, Hussain (1964, pp. 93-106) found very low elasticities ranging from 0.03 to 0.05 in the case of aus rice and aman rice (autumn and winter crops) combined and from 0.09 to 0.13 for aus rice alone.

13. It is instructive to note the insights offered by Smith (1937, p. 159) on the incentives for tenancy in areas of rice culture: "Though its [rice] cultivation requires more labour, a much greater surplus remains after maintaining all that labour.... The real value of his [landlord's] rent, his real power and authority, his command of the necessaries and conveniences of life with which labour of other people could supply him, would necessarily be much greater."

14. The investigators stayed in the villages throughout the year and recorded the data for each agricultural operation pertaining to the selected farms.

15. This measure of profit includes the remuneration for management. However, since the latter can be expected to vary roughly in proportion to the size of the holding and since profits have been expressed on a per-acre basis, its inclusion is unlikely to have vitiated the comparison of variance in profit per acre between zones.

16. Even at the aggregative level, such as among different States of India, area under sharecropping in 1960-61 as a proportion to operated area as well as to tenanted area bears an inverse relationship to uncertainty as indicated by the coefficient of variation in rainfall during 1950-62. The respective correlation coefficients (R) are -.47 (significant at 5 per cent level) and -.50 (significant at 10 per cent level). For the data on area under sharecropping, see Government of India (1966); for the data on the coefficient of variation in rainfall, see Rao (1968). Sharma's study (1965), based on the census of 1961, reveals a similar relationship between the extent of sharecropping and uncertainty.

17. In view of their limited capacity for uncertainty bearing, it is to be expected that small landowners and landless labourers would opt for fixed earnings by hiring out their services in situations of high uncertainty and would prefer to lease in land on a share-rent basis in situations of relative certainty. They may even be willing to pay fixed rents if the variance in yields is negligible, which might well be the case with fixed-kind rents in the rice zone.

18. Marshall (1956, p. 645) explains that share tenancy may be advantageous when the holdings are very small and "is not suitable for holdings large enough to give scope to the enterprise of an able and responsible tenant." The hypothesis proposed in this paper emphasizes how the particular tenurial arrangements as well as the size of holdings may themselves be influenced by the scope for enterprise.

19. The study could not be extended to the year 1959-60, owing to certain gaps in the available data.

20. Input-output data are not available separately for the share-rented portions held by part tenants.

21. Output variable is measured in rupees per farm, prices used for the valuation of crops being the same for all farms. Land input is in standard acres corrected for inter-farm and intervillage differences in fertility as reflected in the respective rental values. Labour variable is represented by the expenses of hired labour plus the imputed value of own-labour services. Labour variable so constructed takes account of the possible differences in its quality. Capital variable consists of the values of seeds, manures, and fertilisers, irrigation charges, and maintenance expenses together with depreciation on fixed equipment, for example, bullocks and implements.

22. A high intercorrelation between labour and land inputs can be expected in certain situations, even when the production functions allow for flexible input combinations. The difference in the allocation of labour between share-rented and owner-operated farms of similar size may not be significant if the sharecroppers' marginal utility of income is higher and if they do not have alternative employment opportunities. That the labour input induced by the tenant's higher marginal utility of income may compensate for the tendency to restrict this input under sharecropping has been noted by Johnson (1950) and Sen (1966). However, the net effect of these opposing tendencies on the allocation of labour cannot be established *a priori* without appeal to the relevant facts (Robbins 1930).

23. Marginal productivity of land (x) is given by by/x, where the value of x is taken at its geometric mean, y is the estimated level of output when x is held at its geometric mean, and b is the elasticity of output with respect to factor x.

24. Heady's study (1955), for instance, represents this approach. Heady also observes that the amounts of capital and labour inputs as well as output per unit of land are lower among share-rented farms than among cash-rented farms. However, as noted in the previous section, if cash rents prevail in situations of high uncertainty and crop sharing in situations characterised by limited scope for decision making, then the input mix may be expected to be different owing to the differing technological conditions, so that such comparisons cannot yield valid inferences regarding resource-use efficiency.

25. It may be asked why, in a free institutional setting, the marginal product of land fails to be equalised between different holdings either through greater renting of land by small farmers or through greater application of inputs per acre among large farms. Raj (1968) has argued that, owing to the production uncertainties, the area that a tenant can rent is limited by the size of his own holding and that larger farms cannot hire the desired amount of labour owing to managerial limitations as well as the "uncertainties of hiring labour." For a theoretical discussion of how the size of own capital invested may limit the capacity of the firm to expand because of uncertainty and capital rationing, see Kalecki (1937).

26. It is possible that, owing to lower investable surplus and capital rationing by credit institutions, sharecroppers would be found to invest less in the modern inputs even when the landlords are prepared to share the costs. That these lower investments cannot be attributed to the sharecropping system as such is evident from a study by Hameed (in preparation) relating to the Thanjavur district where even the tenants paying fixed-kind rents were found to invest less in fertilisers compared with owner-operators of similar size.

27. Marshall (1956, p. 645) noted the existence as well as the usefulness of the practice of sharing inputs in the United States and France. In India, a study of the package districts shows that in the Ludhiana district of Punjab, "the landlord gets 50 per cent of the gross produce" as rent and, in regard to the rate of utilisation of fertilisers, sharecroppers "do not seem to make much distinction between owned land and tenancy land. Where rent is paid as a share of produce, the cost of fertiliser is shared half and half between the owner and the tenant" (Ladejinsky 1965, p. 35). It can be shown that the allocation of resources would be optimal under such situations, that is, when costs on variable inputs are shared by the landlord and the tenant in the same proportion as output is shared. Sen and Varghese (1966) and Adams and Rask (1968) have argued that, under certain market situations and with certain production functions, it may not be profitable for landlords to share the costs on variable inputs. Their argument assumes that sharecroppers will use these inputs up to the point where the marginal product net of rent share is equal to the price. But in view of the greater investable capacity of the landlords, the above argument can be employed to yield the opposite and more plausible result that, in the case of variable inputs like fertilisers, cost sharing may not be profitable for the sharecroppers when they know that the landlords would pay for these inputs up to the point where the marginal product net of tenant's share is equal to the price. However, it is possible to conceive of bargaining situations when cost sharing would prove profitable to both of the parties, for example, when each party is confronted with the possibility that the other party will not pay for fertilisers at all unless the costs are shared. Indeed, unlike the fixed-rent system, sharecropping permits landlords to benefit by promoting the use of fertilisers through cost sharing in situations where the sharecroppers may not otherwise apply such inputs owing to lack of knowledge, shortage of capital, risk aversion, etc.

28. A recent study (Day 1967) of agricultural transformation in the Mississippi Delta of the United States indicates that the creation of profitable investment opportunities through the diffusion of new capital-intensive techniques led to the demise of the sharecropper.

References

Adams, D.W., and Rask, N. (1968), "Economies of Cost-Share Leases in Less-developed Countries," *American J. Agricultural Economics* 50, No. 4 (November): pp. 935-42.

Alchian, A.A. (1950), "Uncertainty, Evolution, and Economic Theory," *J.P.E.* 58 (June): pp. 211-21.

Arrow, K.J. (1951), "Alternative Approaches to the Theory of Choice in Risk-taking Situations," *Econometrica* 19 (October): pp. 404-37.

Bhagwati, J., and Chakravarty, S. (1969), "Contributions to Indian Economic Analysis: A Survey," *A.E.R.* 59 (suppl. September): pp. 1-118.

Cheung, Steven N.S. (1969), *The Theory of Share Tenancy*. Chicago: University Chicago Press.

Day, R.H. (1967), "The Economics of Technological Change and the Demise of the Sharecropper," *A.E.R.* 57 (June): pp. 427-49.

Government of India, (1957), *Studies in Economics of Farm Management: West Bengal*, New Delhi: Directorate Econ. and Statis.

———. (1963), *Studies in Economics of Farm Management: West Godavari (Andhra), 1957-58 to 1959-60*, New Delhi: Directorate Econ. and Statis., Mimeographed.

———. (1966), *National Sample Survey, Sixteenth and Seventeenth Round, Number 177*, New Delhi: Indian Statistical Institute, Mimeographed.

Hameed, N.D.A. *Prospects of Increasing Agricultural Production through Intensive Cultivation — A Case Study of the Rice Economy of Thanjavur District, India*, Ph.D. dissertation, Univ. Delhi, in preparation.

Heady, E.O. (1955), "Marginal Resource Productivity and Imputation of Shares for a Sample of Rented Farms," *J.P.E.* 63 (December): pp. 500-511.

Hussain, S.M. (1964), "A Note on Farmer Response to Price in East Pakistan," *Pakistan Development Rev. 4* (Spring): pp. 93-106.

Johnson, D. Gale, (1950), "Resource Allocation under Share Contracts," *J.P.E. 58* (April): pp. 111-23.

Kalecki, M. (1937), "The Principle of Increasing Risk," *Economica,* n.s., 4, No. 16 (November): pp. 440-47.

Knight, Frank H. (1965), *Risk, Uncertainty, and Profit.* New York: Harper & Row.

Krishna, R. (1963), "Acreage Response Functions for Some Punjab Crops," *Econ. J.* 73 (September): pp. 477-87.

Ladejinsky, W. (1965), *A Study on Tenurial Conditions in Package Districts*, New Delhi: Planning Commission.

Marshall, Alfred, (1956), *Principles of Economics*, London: Macmillan.

Narain, D. (1965), *Impact of Price Movements on Areas under Selected Crops in India, 1900-1939*, Cambridge: Cambridge University Press.

Nerlove, Marc, (1958), *The Dynamics of Supply: Estimation of Farmers' Response to Price*, Baltimore: Johns Hopkins Press.

Raj, K.N. (1968), *Accumulation and Growth in Agrarian Economies*, Mimeographed; available from author.

Rao, C.H.H. (1966a), *Taxation of Agricultural Land in Andhra Pradesh*, Bombay: Asia Publishing House.

————. (1966b), "Alternative Explanations of the Inverse Relationship between Farm Size and Output per Acre in India," *Indian Econ. Rev.*, n.s., 1, No. 2 (October): pp. 1-12.

————. (1968), "Fluctuations in Agricultural Growth: An Analysis of Unstable Increase in Productivity," *Economic and Political Weekly 3* (January): pp. 87-94.

Robbins, L. (1930), "On the Elasticity of Demand for Income in Terms of Effort," *Economica* 10 (June): pp. 123-29.

Robinson, E.A.G. (1962), *Structure of Competitive Industry*, Chicago: University of Chicago Press.

Schultz, T.W. (1940), "Capital Rationing, Uncertainty, and Farm Tenancy Reform," *J.P.E.* 48 (June): pp. 309-24.

———. (1965), *Transforming Traditional Agriculture*. New Haven, Conn.: Yale Univ. Press.

———. (1968), "Institutions and the Rising Economic Value of Man," Paper presented at the *American Agricultural Economies Association Meeting*, August.

Sen, A.K. (1966), "Peasants and Dualism with or without Surplus Labour," *Journal of Political Economy 74* (October): pp. 425-50.

Sen, A.K., and Varghese, T.C. (1966), "Tenancy and Resource Allocation," *Seminar* (India) no. 81 (May), pp. 28-33.

Sharma, P.S. (1965), "A Study of the Structural and Tenurial Aspects of Rural Economy in the Light of 1961 Census," *Indian J. Agricultural Econ. 20* (October-December): pp. 46-82.

Shickele, Rainer, (1941), "Effect of Tenure Systems on Agricultural Efficiency," *J. Farm Econ. 23* (February): pp. 185-207.

Smith, Adam, (1937), *The Wealth of Nations*, New York: Modern Library.

Venkataraman, L.S. (1958), *A Statistical Study of Indian Jute Production and Marketing with Special Reference to Foreign Demand*, Ph.D. dissertation, University of Chicago.

Reprinted from *The Journal of Political Economy*, Vol.79, No.3, May/June 1971.

21

Farm Size and Yield Per Acre: A Comment

In his article "Farm Size and Yield Per Acre" (*Economic and Political Weekly*, Special Number, July 1968) Ashok Rudra claims that his findings, based on the data pertaining to 20 villages collected by the Agro-Economic Research Centre, University of Delhi, "flatly contradict the earlier finding of the Farm Management Studies that yield per acre declines with the increase in farm size". Rudra suspects that the process of aggregation is responsible for what he calls the "spurious correlation" of the Farm Management Studies. He observes that "if they had only looked at the nature of the scatter of the disaggregated raw data they could not have drawn the conclusions that were drawn on the basis of aggregated data representing size group averages for entire districts".

Different Concept of Yield

Rudra is right in asserting the need to examine the disaggregated data on size of holding and yield per acre for each village separately before any attempt is made to establish the correlation between the two. However, it can be shown that the divergence between his findings and those of the FMS would persist despite disaggregation. This is because his concept of "yield per acre" is substantially different from that of FMS and is in fact biased towards eliminating the inverse relationship between size of holding and yield per acre.

Rudra's definition of the size of a farm ("the total of net cultivated area and uncultivated area") is the same as used in FMS. But whereas yield per acre in FMS is obtained by dividing the gross value of output by the *size of farm* as defined above, Rudra obtains yield per acre "by dividing the gross value of output by the *gross cultivated area*". Rudra tries to justify his procedure by suggesting that one should "treat the extent of double cropping, proportion of uncultivated land to cultivated land, etc., as distinct factors affecting the economics of farming and not subsume all of them in the factor 'size of farm'".

Factors like the extent of double cropping and the proportion of uncultivated land to cultivated land reflect the economics of farming instead of being exogenous factors affecting the economics of farming. A farmer cultivating his land more intensively through multiple cropping and by raising the proportion of cultivated land to uncultivated land obtains higher yield per acre from his holding than the one who for one reason or the other fails to do so. In such a case, the former can be called a more efficient user of land than the latter. The whole question of the relationship between the size of holding and productivity assumes significance for economic analysis only insofar as it throws light on the efficiency with which land is used among different size groups of holdings. Rudra's procedure, in effect, corrects for the efficiency of land use because his divisor is gross cropped area. This point is illustrated below with the help of hypothetical figures for two farms.

		Farm A	*Farm B*
(1)	Size of holding (acres)	10	10
(2)	Output per farm (Rs.)	750	500
(3)	Output per acre held (row 2/row 1)	75	50
(4)	Percentage of holding cultivated	100	80
(5)	Percentage of cultivated land double cropped	50	25
(6)	Gross cropped area (acres)	15	10
(7)	Output per acre of gross cropped area (Rs.) (row 2/row 6)	50	50

In the above illustration farms A and B with identical holdings show differences in efficiency. The output of farm A is higher than that of B because of the higher proportion of cultivated area (row 4) as well as the greater intensity of cropping (row 5). Output per acre held is therefore higher for A than for B (row 3) indicating that A is using his land more efficiently than B. But if one follows Rudra's procedure and divides output by gross cropped area, which is higher for A than for B (row 6), output per acre turns out to be the same (row 7) indicating that both are equally efficient.

It is now easy to understand why Rudra could not find any inverse relationship between the size of holding and yield per acre, if it is assumed that the relationship between the intensity of land use (*i.e.*, multiple cropping, *etc.*) and size of holding as revealed by the FMS holds true for the villages he studied. The FMS have revealed that in all the districts studied, the percentage of cultivated to uncultivated area as well as the percentage of cultivated area cropped more than once decrease sharply with increase in the size of holding. This is because, among other factors, the percentage of holding irrigated invariably declines with increase in the size of holding. These factors give rise to the inverse relationship between farm size and output per acre held. Thus if Rudra's procedure is followed for analysing FMS data, then one would be correcting for the efficiency factors mentioned above and it is very likely that one might get the kind of picture Rudra got.

Corresponding Inverse Relationship

On the other hand, if the villages under study are such that the intensity of cultivation does not vary inversely with the size of holding as in FMS, then it is possible that no relationship between farm size and yield per acre will be found, even if the concept of yield per acre is the same as in FMS. A.P. Rao's study ("Size of Holding and Productivity", *Economic and*

Political Weekly, November 11, 1967), which Rudra quotes in his support, represents such a case. Of the five villages selected for study by Rao, two are completely irrigated, two completely unirrigated and in the remaining one the percentage of holding irrigated is constant for different size groups. That Rao did not find any inverse relationship between farm size and yield per acre should not, therefore, be surprising. His study only confirms that the inverse relationship of the FMS between size of holding and output per acre arises from the corresponding inverse relationship regarding intensity of cropping, irrigation, *etc*. Therefore, the relevant question in this case is not of methodology but whether these five villages are to be regarded as more typical of Indian farming than those studied by FMS in different parts of the country.

As for Rudra's analysis, one hopes that in his promised scrutiny of FMS data he tests the relationship between size of holding and yield per acre by following the concept of yield per acre adopted by FMS, which is more meaningful for the economic analysis of the efficiency of land use. Also, one hopes that he provides a new set of results by using this concept to test the relationship between size of holding and yield per acre in the twenty villages studied by him.

Economic and Political Weekly, September 14, 1968.

22

Farm Size and Credit Policy

Land and labour have ceased to be the predominant factors of growth in Indian agriculture. Capital, together with scientific knowledge, has already become a major source of growth, and its significance is rapidly increasing. The disparities in income and in wealth, within the agricultural sector, would therefore depend crucially on how capital is and will be distributed among the different holding-size groups. Those committed to an egalitarian agrarian set-up have been focusing attention almost exclusively on the need to have a more equal distribution of land, neglecting the equally important issue of equitable distribution of capital. Credit from institutional sources (*e.g.*, cooperatives, government, and the nationalised and commercial banks) is becoming increasingly important as a means of capital formation in agriculture. These sources accounted for about one-third of total agricultural credit at the beginning of the Fourth Plan, as against 7.3 per cent in 1951-52.

Creditworthiness

In a capitalist or free market economy, the relative growth of different firms may depend essentially on their initial wealth position[1] — because, among other things, the ability of a firm to borrow on reasonable terms in the market depends on its 'creditworthiness'. It is possible to argue that since large farmers save a greater proportion of the increment in their income than do small farmers, they would add to their wealth at a faster rate and would eventually proletarianise the small farmers.[2] A qualification to this is provided, in principle at least, by the public regulation of the capital markets. Given a favourable socio-political environment, a public authority can, through its own operations in the capital market, make capital cheaper and more accessible to the small farmers and so improve their competitive position *vis-à-vis* the large farmers. It would be interesting, therefore, to evaluate the performance of institutional credit in this respect and to explore the possibilities of improving its effectiveness in the future.

Rising Significance of Capital

Between 1960-61 and 1968-69, agricultural production (all commodities) rose by 11.6 per cent, or at the simple rate of about 1.5 per cent per annum, while cropped area increased by only about 0.5 per cent per annum.[3] Although the country's population may have been growing at about 2.5 per cent per annum, the annual growth of employment in agriculture is unlikely to have exceeded 2 per cent. Assuming that, a one per cent rise in land input increases output by 0.3 per cent, other factors remaining constant, the growth of cropped area accounts for only about 10 per cent of the growth in output in this period. If we take a similar 'output response' with respect to the labour input,[4] the contribution of labour is about 40 per cent. Thus the growth of capital and knowledge together accounted for the remaining 50 per cent of the growth in output. In fact, for the periods 1960-61 to 1967-68 and 1961-62 to 1967-68, the contribution of capital amounts to 55 per cent and 59 per cent respectively.

Distribution of Credit

In the volume of credit supplied, the institutional sources, as compared to private sources, have proved to be discriminative against the small farmers. According to the All India Rural Credit Survey, in 1951-52, small farmers showed higher indebtedness per acre than did large farmers. But credit per acre supplied by co-operatives to small farmers was significantly lower than that for large farmers. This implies that credit per acre from private sources was higher for small farmers. The tendency has continued through the late sixties. In the second group of IADP districts, for instance, loan per hectare from all sources in 1967-68 was Rs. 67 and Rs. 54 for small and large farmers, respectively. Of this, the amounts advanced by co-operatives were Rs. 30 and Rs. 33, respectively.[5]

According to the All India Rural Debt and Investment Survey (1961-62), the distribution of the credit from the co-operatives was inequitous as among different asset groups (Table 22.1). Although productive assets (*e.g.*, irrigation and cattle) per acre decline with increase in size of holding,[6] credit per acre extended by the co-operatives is found to increase significantly with increase in operated area (see columns 5 and 6 of Table 22.1). Among the various States, wherever concentration of assets is higher, concentration of co-operative credit is also higher[7] indicating that the political power of the large farmers, which derives largely from the disparities in wealth, is an important factor determining the distribution of co-operative credit.

A study made by the Programme Evaluation Organisation of the Planning Commission on the working of 84 small and 33 large co-operative credit societies in the country revealed that, during all the three years covered by the Survey, in the case of the small societies, the percentage of overdues to loans outstanding was significantly higher among large farmers as compared to among small farmers. Among large farmers, the overdues constitute as much as 30 to 46 per cent of the loans outstanding (Table 22.2).

Table 22.1

Relative Importance of Co-operatives in the Total Borrowings of Cultivators among Different Asset Groups, 1961-62

Asset Group (Rs.)	*Percentage of Families Borrowing from Co-operatives*	*Percentage of Those Borrowing from Co-operatives to Those Borrowing from All Agencies*	*Percentage of Borrowings from Co-operatives Accounted by Different Assets Group*	*Percentage of Amount Borrowed from Co-operatives to Total Borrowings*	*Average Area Personally Cultivated Per Cultivator Household (Acre)*	*Average Borrowings Per Acre (Rs.)*
(1)	*(2)*	*(3)*	*(4)*	*(5)*	*(6)*	*(7)*
Less than 500	2.6	5.7	0.5	5.1	1.27	1.81
500-1000	3.9	8.2	1.5	7.3	2.14	2.38
1000-2500	6.2	12.7	7.4	9.6	3.35	2.81
2500-5000	9.4	17.6	14.3	12.6	5.35	3.66
5000-10000	12.3	22.0	20.7	14.2	8.06	4.42
10000-20000	15.4	27.6	21.4	16.5	11.86	5.36
20000 and above	19.0	34.9	34.2	23.1	23.26	7.33
All asset groups	**9.4**	**18.1**	**100.0**	**15.5**	**6.63**	**4.81**

Source: All-India Rural Debt and Investment Survey: quoted from the Report of the All-India Rural Credit Review Committee, pp. 127-28.

In the case of large societies, however, percentage of overdues to loans outstanding does not show any distinct tendency in relation to the size of holding. Further, the small farmers (with holdings size below 10 acres) get a better share in the credit advanced from large societies than from small societies. These facts indicate that large farmers, owing to their greater bargaining power at the local level, are able to corner a substantial proportion of the credit from the small societies. Also, they are able to get away with large overdues. Between the large and small societies, the small farmers seem to get a better deal from the large societies.

Table 22.2

Advances According to Size of Holding by Large and Small Societies

Size of Holdings (acres)	*Large Societies*		*Small Societies*	
	Percentage Distribution of Loans Advanced	*Percentage Overdues to Loans Outstanding*	*Percentage Distribution of Loans Advanced*	*Percentage Overdues to Loans Outstanding*
1956:				
Below 2	7.8	13.0	7.4	17.2
2 - 5	25.6	12.2	21.5	9.9
5 - 10	22.5	12.4	20.6	16.4
10 - 25	24.0	0.0	27.0	22.4
Above 25	20.1	..	23.5	40.5
All groups	*100.0*		*100.0*	
1957:				
Below 2	15.5	4.7	9.5	19.9
2 - 5	27.8	3.1	24.7	13.5
5 - 10	24.5	6.3	22.4	17.1
10 - 25	21.8	4.3	27.8	20.3
Above 25	10.4	2.3	15.6	46.4
All groups	*100.0*		*100.0*	
1958:				
Below 2	8.6	10.0	8.5	9.8
2 - 5	21.9	8.8	21.4	9.1
5 - 10	19.8	14.6	20.9	10.8
10 - 25	29.1	8.3	30.1	11.4
above 25	20.6	11.7	19.1	31.2
All groups	*100.0*		*100.0*	

Source: Programme Evaluation Organisation, Planning Commission, 'Evaluation Report On the Working of the Large and Small Sized Co-operative Societies', April 1959, pp. 52-56.

Credit Rationing

In a situation where large farmers wield considerable power at the local level, it would be impossible to ensure an equitable distribution of institutional credit without a clear-cut policy from above regarding the allocation of credit among the different land holding groups. Such a policy would also have to be backed by measures to enforce such limits. We need to practise credit rationing if institutional credit is not to become a means for widening the existing disparities in income and wealth within the agricultural sector — especially if it is indeed to be made an instrument for reducing such disparities. Co-operatives and commercial and nationalised banks should be required to allocate to small farmers — say those with holdings below 10 acres — a specified quantum which should be more than proportionate to the land held by such farmers in the region concerned. This is because, as mentioned above, small farmers in general have larger productive assets per acre than large farmers. Besides, large farmers are in a better position to finance investment from their own resources, Central and State Governments and the Reserve Bank can devise various methods to ensure compliance with such quotas.

Such a rationing would discourage consumption expenditure by large farmers, restrict the scope for re-lending of the credit taken from co-operatives, and would thus reduce the volume of overdues. Insofar as credit availability is increased significantly for the small farmers, their dependence on agricultural moneylenders would decline. Since a good part of the demand for credit by small farmers is therefore inelastic,[8] increased availability of credit to them may result in lower interest rates being charged by the agricultural moneylenders (Table 22.3 illustrates how the amounts borrowed at higher rates of interest increase with the decline in the size of holding). The agricultural moneylenders may thus find it more attractive, at the relevant margin, to increase their deposits with the banks. Credit rationing would thus also contribute to the economic viability of credit institutions.

What would be the impact of such a redistribution of credit on agricultural productivity and surplus? Reallocation of credit in favour of small farmers would, to a large extent, mean substitution of institutional sources for other sources. To this extent, the supply of credit on better terms represents the only

Table 22.3

Percentages of Total Amount Borrowed at Different Rates of Interest According to Size of Holding in Marupaka Village, Karimnagar District

Size of Holding	*At 12 Per cent*	*At 18 Per cent*	*At 24 Per cent*	*Without Any Interest*	*Total*
Below 5 Acre	3567 (43.72)	3778 (46.30)	700 (8.58)	114 (1.40)	8159 (100)
5 to 10 Acre	3190 (76.85)	401 (9.66)	560 (13.49)	— —	4151 (100)
10 to 20 Acre	1440 (100)	—	—	—	1440 (100)
20 to 50 Acre	600 (100)	—	—	—	600 (100)
50 to 100 Acre	—	—	—	—	—
Total	**8797 (61.30)**	**4179 (29.12)**	**1260 (8.78)**	**114 (0.80)**	**14350 (100)**

Note: Figures in brackets are percentages.

Source: C.H. Hanumantha Rao, "Agricultural Marketing in Karimnagar District", 1956, (Mimeographed).

gain for the small farmers. Therefore, the prospect that a good part of institutional credit is likely to be spent for consumption will not make any difference to the pattern of credit utilisation for the sector as a whole. Further, to the extent this reallocation leads to a net transfer of credit (all sources) from the larger to the small farms for productive purposes, there is most likely to be an increase in the total agricultural output. This is because, owing to the greater availability of labour and other complementary inputs, such as irrigation, among small farms, a rupee spent on current inputs, say fertilisers, brings forth significantly larger returns than on large farms.[9] Since investments in modern inputs are highly profitable, it should be possible for the small farmers to repay the loans in time and still be left with an increase in their net income.

Although small farmers consume a larger proportion of the increase in their income than the large farmers, the welfare implications of this redistribution cannot be overlooked: The rural rich too have a high propensity to consume. The upper income groups in the rural areas seem to consume at least 82 paise from a rupee increase in their income.[10] This means that, the transfer of productive resources from larger to smaller farmers has a substantial potential for redistributing consumption.

The prevailing rates of interest charged by the institutional agencies for agricultural loans are only slightly lower than for the non-agricultural sector. This marginal differential in favour of agriculture may not be without justification. For one thing, Indian agriculture is still subject to high production uncertainties so that farmers discount the returns significantly while making investments. Since the process of agricultural modernisation in India is still subject to several uncertainties, the existing incentive in the form of slightly lower interest rates may help to accelerate the modernising process. Besides, institutional credit accounts for only about one-third of total agricultural credit and the interest rates charged by other sources vary from 12 to 36 per cent; consequently, the effective average (weighted) interest rate for the farmers is unlikely to be lower than 18 per cent — which may be much higher than the average for the non-agricultural sector.

It may be argued that a higher interest rate on borrowings by large farmers coupled with a higher rate on deposits may improve the competitive position of institutional agencies in the rural capital market. Such a policy may, no doubt, discourage diversion of loans for consumption by large farmers. But since investments cannot be considered to be insensitive to interest rates, genuine productive investments which may account for at least 50 per cent of borrowings among large farmers (assuming that overdues represent bad loans) may also be discouraged. Since there is at present no economic basis for discriminating against investments in agriculture *vis-a-vis* industry, such a policy may prove harmful for the economy insofar as it leads to imbalances in growth. The insufficiency of investments in agriculture is evident from the very high marginal returns at present, as the latter have to be equated with the prevailing high (average) interest rates after discounting for risk.[11] For the same reasons, higher interest rates charged to large farmers would be inefficient as a measure of equity compared to direct taxation of their incomes. Upper agricultural income groups should be subjected to income tax and other direct taxes in the same way that their urban counterparts are taxed.

True, the prevailing high interest rates in the free market provide an incentive for large farmers to re-lend a good part of the borrowings from the institutional sources — which may be an important explanation of the high overdues. But so long as large farmers corner a substantial proportion of the credit and small farmers are denied their due share from institutional sources, the latter's dependence on free market will not be reduced and it would be possible for the agricultural moneylenders (large farmers) to shift the burden of high interest rates to the small farmers by raising market rates since the demand for such loan is inelastic. The rising market interest rates and the continuance of small farmers as a profitable venue for lending may neutralise the effect of a rise in deposit rates. Thus there may not be enough incentive for agricultural moneylenders to increase their deposits with the banks.

At present, the real costs of borrowing from institutional sources may be very high for small farmers. They may have to pay several dues to intermediaries, fulfil cumbersome formalities, and even then credit may not be available in time. This may, indeed, explain why small farmers have a low share in the credit from institutional sources. Credit rationing, insofar as it increases availability of credit to them may reduce the cost of obtaining loans. The scope for reducing non-interest costs may be so significant that, if attention is focused on them, reduction of interest rates may not become necessary as a measure for attracting small farmers to institutional sources.

It is evident, then, that differential interest rates cannot be a substitute for credit rationing. On introducing credit rationing, they may prove unnecessary as instruments to discourage consumption expenditure and re-lending, and they may even have disincentive effects on genuine investments. Since reallocation of credit will take time, and interest rates come into force immediately, the impact of higher interest rates may not prove beneficial on balance. Lower interest rates on smaller loans, on the other hand, would reduce the cost of obtaining loans and this may be essential for attracting small farmers till such time as credit rationing becomes fully effective. Considering the fact that the market rate facing them is quite high, the average (weighted) interest rate, even after a small reduction on institutional loans, may still be higher when compared to the average rate for large farmers. Assuming that small farmers will get about 50 per cent of institutional credit (out of about Rs. 600 crore), a reduction in the interest rates for them by 2 per cent would cost about Rs. 6 crore per annum. This should be regarded as a subsidy borne by the community which can be more than compensated by other progressive measures such as increased taxation of rich farmers.

Creditworthiness of Small Farmers

As indicated earlier, the bargaining position of the small farmers in the credit market cannot be strengthened without meeting a part of their loan requirements for pressing consumption. There is no basis for regarding such loans as unproductive, as they constitute working capital and are self-liquidating.[12] In fact, a significant proportion of borrowing by large farmers for "current agricultural expenses" is used for the payment of wages.

However, the position of small farmers can be improved only by raising their creditworthiness. Credit policy should be such as to generate more assets which are of the self-liquidating type, like irrigation and dairy cattle, and should also encourage techniques which are land-saving and labour-using. Creditworthiness can also be increased by encouraging a production pattern that is market-oriented and biased against self-consumption.[13] In the choice of investment, the balance can be tilted in favour of insurable assets and against the tendency for sizeable investment in inputs even where the element of risk is very high.

An effective way of improving the viability of small farms is to reduce their loan requirements for long-term investments *e.g.*, draft cattle. This would also render their enterprise more risk-free. The demand for tractors in Indian agriculture derives to a significant extent from the high cost of bullock labour.[14] Small farmers can be helped to benefit from the use of tractors if tractor services are hired out to them by institutions such as agro-industrial corporations. The pattern of tractorisation has been almost exclusively in favour of large farmers which enables them not only to dispense with the high costs on bullocks and so to make their enterprise less risky but also gives them considerable power for charging higher rates to the small and middle farmers. The extent of possible advantage and power, derived by the large farmers, is evident from the fact that the number of tractors almost trebled from 31,000 to 91,000 — within a period of 8 years from 1960-61 to 1968-69.[15] Credit policy pursued so far is obviously responsible for this concentration of assets and economic power. A major shift in the concentration of assets can be effected if preference is given to public institutions while extending tractor loans. This is especially important because, unlike pump-sets for irrigation, private ownership of tractors does not seem to be essential for the efficient management of modern agriculture. Public ownership of tractors can also help to provide employment to landless labourers as drivers, instead of providing such opportunities to the rich farmers, who prefer tractor driving to ploughing with bullocks.

Another major step for improving the assets position of the small farmers *vis-à-vis* the larger land-holders would be to extend long-term interest-free loans to small farmers, say those with holdings below 5 or 3 acre, for buying land in the market.[16] Such loans may be linked with purchases of land from big land-holders — *i.e.*, holders of land above 25 acres.

Many of the measures to generate the relatively risk-free and self-liquidating assets, by encouraging enterprises like dairying, poultry farming and piggary *etc.*, would involve opening up of urban markets through investments in roads and communications. Linking up credit with marketing could be postponed till now because of the predominance of the large farmer as loan receiver. But small farmers cannot be made viable unless credit is linked with marketing. That would involve the provision of storage and marketing services on a significant scale.

Conclusion

Whether institutional credit will be made an instrument for reducing the disparities in income and wealth in the agricultural sector will finally depend on the awareness and bargaining power of the small and middle farmers themselves. At present, their interests do not seem to offer an attractive source of political power for the parties concerned, for whom the returns from alternative sources continue to be higher. It is necessary, therefore, to start with institutions where small farmers are likely to get a better deal. Paradoxically, at present, the greater the distance at which the institution is located from the village, the better the treatment which the small farmer is likely to get. The initiative from the Central Government and the nationalised banks would, therefore, be most rewarding.

Notes

1. M. Kalecki, 'The Principle of Increasing Risk', *Economica* New Series, IV, No. 16, 1937.
2. For a presentation of this proposition see, Pranab Bardhan, "A Model of Growth of Capitalism in a Dual Agrarian Economy", Indian Statistical Institute, Planning Unit, New Delhi, (Mimeographed).
3. These estimates are based on the All-India Index Numbers of Area and Agricultural Production given in *Agricultural Situation in India*, November 1967 and August 1969.
4. We have made these assumptions regarding the output elasticities after going through the results of several production function studies and noting their limitations.
5. "Modernising Indian Agriculture", Ministry of Agriculture, *Report on the Intensive Agricultural District Programme* (1960-68), Vol. I, p. 109.
6. For the data on irrigation see, "Agricultural Production Functions, Costs and Return", *op. cit.* For the data on the distribution of cattle see C.H. Hanumantha Rao, 'India's 'Surplus' Cattle: Some Empirical Results' *Economic and Political Weekly*, Review of Agriculture, December 1969.
7. About 70 per cent of the inter-State variation in the concentration of credit is explained by the concentration of assets. An increase in the co-efficient of concentration of assets by 0.10000 is associated with an increase in the co-efficient of concentration of credit to the extent of 0.10450 (significant at the one per cent level).
8. K.N. Raj, 'Saving, Capital Transfers and Rates of Interest' in his "Accumulation and Growth in Agrarian Economies", 1968, (Mimeographed). An analysis of the inter-rational data presented by Raj (taken from the Rural Credit Survey) shows that when the percentage of borrowing from agricultural or professional moneylenders for consumption increases by 1 point, the percentage borrowed from them at more than 18 per cent interest rate increases by 0.8 (significant at 5 per cent level; R^2=0.24).
9. According to G.R. Saini's study for Uttar Pradesh and Punjab covering two years (1955-56 and 1956-57), the marginal productivity of a rupee invested in fertilisers ranged between Rs. 2.04 to Rs. 2.44 among small farms as against Rs. 0.99 to Rs. 1.84 among large farms. See his "Resource-Use Efficiency in Agriculture", *The Indian Journal of Agricultural Economics*, April-June 1969.
10. We have estimated this marginal propensity to consume for income class above Rs. 1,200 per annum from the data contained in "All India Rural Household Survey", Volume II, "Income, Investment and Saving" by the National Council of Applied Economic Research. The marginal propensity to consume for the incomes below Rs. 1,200 per annum works out to be 0.97 which is not significantly different from unity.
11. B.M. Desai and D.K. Desai "Is Inadequacy of Institutional Credit a Problem in Changing Agriculture?", *Economic and Political Weekly*, Review of Agriculture, September 26, 1970. See also Saini, *op. cit.*
12. Raj, *op. cit.*
13. For a thoughtful discussion of some of these aspects see C.B. Baker, 'Limited Capital as a Restraint on Agricultural Development', in *Economic Development of Agriculture*, Iowa State University Press, Ames, Iowa.

14. S.V. Sethuraman, *Long Run Demand For Draft Animals in Indian Agriculture*, a Ph.D Dissertation (unpublished), University of Chicago, 1969.

15. Government of India, *Fourth Five-Year Plan*, p. 115.

16. Late M. Kalecki made this suggestion in an article in *The Economic Weekly* after his visit to India in 1959-60.

Economic and Political Weekly, Vol. V, No. 52, Review of Agriculture, December 26, 1970.

23

Differential Rates of Interest: A Comment

In his article "On Differential Rates of Interest" (EPW, July 1, 1972, p. 1279) A.K. Dasgupta observes that although banks should charge differential rates of interest with a view to discriminating in favour of the weaker customers, they should operate on commercial lines and seek to maximise profits subject to some subsidy to the weaker sections. According to him, such an operation on commercial lines is possible because banks have a monopolistic position in the credit market and the system of differential interest rates amounts to monopoly price discrimination, if only re-lending by the weaker sections can be discouraged, as the elasticity of demand for credit of the poorer sections such as small farmers is higher than that of the richer sections.

I shall argue in this note that the theory of monopoly price discrimination is relevant for understanding the capital markets as it operates, especially in the rural sector, and that the high rates of interest charged to the weaker sections by the private moneylenders can be explained in terms of this theory. Contrary to Dasgupta's assumptions, this implies lower elasticity of demand for credit in agricultural sector and among small farmers. Therefore, the case for charging differential rates of interest by the public financial institutions derives not from the theory of monopoly price discrimination but from the theory of price control and rationing, as one of the functions of the public financial institutions is to protect the weaker sections from monopolistic exploitation by the private moneylenders. Thus what should worry the policymakers is not the prospect of relending by the poorer sections but the possibility of relending by the richer to the poorer sections at higher rates of interest if the regime of differential interest rates (prices control) is not combined with the rationing of credit.

Elasticity of Credit Demand

Private moneylenders including agricultural moneylenders have been a dominant source of credit supply in the rural sector, although their relative importance has been declining over the last two decades owing to the entry of institutional sources. Credit supplied from the institutional sources (*e.g.*, co-operatives, government, and the nationalised and commercial banks) formed about one-third of total agricultural credit at the beginning of the Fourth Plan as against 7.3 per cent in 1951-52. Private moneylenders generally charge higher rates of interest to the poorer customers. These higher rates of interest may no doubt be explained, in part, by the higher unit costs of servicing and the risk premium, as the risk of default could be greater in the case of small farmers. But these higher interest rates may also contain an element of monopoly profit which may be greater in the case of small farmers than in the case of richer sections. The possibility of such a monopolistic price discrimination is indicated by the nature of the demand for credit in the rural sector.

The demand for credit is a derived demand because credit is needed to buy inputs in order to produce output for sale. The elasticity of this derived demand for credit is, therefore, governed by the familiar rules of derived demand. Since the elasticity of demand for agricultural commodities is lower than that for non-agricultural goods and services, the elasticity of demand for farm credit can be expected to be lower than that in the non-farm sector. As between the large and small farmers, the elasticity of demand for credit can be expected to be lower in the case of the latter for a variety of reasons. In the first place, large farmers can supplement borrowing with own savings to a considerable extent whereas small farmers have to depend almost entirely on moneylenders. Secondly, the proportion of fixed costs, including family labour, which can not be readily transferred to the alternative occupations is much larger in the case of small farms. Thirdly — and this derives from the second — the demand for credit for meeting the immediate consumption needs is greater in the case of small farmers. The consumption loans in this case have to be regarded as working capital because a good part of the borrowings by the large farmers are used for the payment of wages to the hired labour. These factors may explain, in part, why the rates of interest charged are higher in the farm sector than in the non-farm sector and within the farm sector why those charged to the poor farmers are higher still.[1]

Since monopoly price discrimination could be associated with the restriction of supply to the weaker sections, it is possible that the existing availability of credit is 'insufficient' in the agricultural sector, particularly for small farmers. This seems to be evident from the high marginal productivity of farm credit at present, as the latter has to be equated with the prevailing high interest rates.[2] Similarly, there is evidence to suggest that owing to the greater availability of fixed resources such as family labour among small farms, a rupee spent on current inputs like fertilisers yields significantly larger returns than on large farms.[3]

The expansion of credit to the agricultural sector, particularly to the small farmers, by the public financial institutions at concessional rates of interest can be regarded as a necessary measure to counter the monopolistic tendencies in the capital markets. Since higher rates of interest charged to the large farmers by the institutional sources cannot exceed the marginal returns to investment and since the marginal productivity of capital among small farms seems to be higher when compared to that among large farms, there would be little gain for the small farmers in relending money (borrowed at concessional rates of interest) to the large farmers, as the net return from investment on their own farms would be well above the possible gain from relending. Moreover, institutional credit constitutes only a fraction of the total borrowings of small farmers and rates of interest on borrowings from private sources exceed the 'higher' interest rates to be charged to the richer sections by the institutional sources. Therefore, poorer sections stand to gain by substituting institutional credit for credit from private sources instead of relending to the richer sections.

Since credit is in short supply and since the demand for credit of the poorer sections is relatively inelastic resulting in the high rates of interest charged to them by the private sources, there is always an incentive for the bigger farmers to corner a substantial proportion of institutional credit, even if it is at higher interest rates, and relend it to the small farmers by shifting the burden of high interest rates to them. There is enough evidence to suggest that large farmers, owing to their political power, are able to corner a disproportionately large volume of credit from the institutional sources.[4] It is not without significance, therefore, that the share of agricultural moneylenders (rich farmers) in the total credit advanced in the rural sector has been progressively increasing over the last two decades. Therefore, the policy of differential rates of interest, if it is to fulfil its objective, has to be combined with the rationing of credit with a view to reducing the dependence of the weaker sections on the private sources.[5] Co-operatives and commercial and nationalised banks should be required to allocate to weaker sections a specified proportion of the total credit advanced.[6] Such rationing would restrict the scope for relending of credit taken from institutional sources by the richer sections. Insofar as credit availability is increased significantly for weaker sections, their dependence on money-lenders would be reduced. The demand for credit of the weaker sections being relatively inelastic the increased availability of credit may result in lowering significantly the rates of interest charged to them by the money-lenders.

Notes

1. See in this connection K.N. Raj "Saving, Capital Transfers and Rates of Interest" in his *Accumulation and Growth in Agrarian Economies*, 1968 (mimeographed) and C.H. Hanumantha Rao, "Farm Size and Credit Policy," *Economic and Political Weekly*, Review of Agriculture, December 26, 1970.
2. B.M. Desai and D.K. Desai, "Is Inadequacy of Institutional Credit a Problem in Changing Agriculture?" *Economic and Political Weekly*, Review of Agriculture, September 26, 1970.
3. G.R. Saini, "Resource-Use Efficiency in Agriculture," *The Indian Journal of Agricultural Economics*, April-June 1969.
4. C.H. Hanumantha Rao, "Farm Size and Credit Policy," *op cit.*
5. See the dissent note by Ashok Mitra to the Report of the Committee on Different Interest Rates, Reserve Bank of India, 1971.
6. The Government of Andhra Pradesh is reported to have directed its Primary Agricultural Co-operative Credit Societies to disburse to the small farmers fifty per cent of the total amounts borrowed by them from the co-operative financial institutions for granting short or medium term loans to their members.

Economic and Political Weekly, Vol. VII, No. 37, September 9, 1972.

24

Small and Marginal Farmers in the Context of Economic Reforms

When agriculture can be carried on by large farmers, by medium farmers, by corporations, then why should we particularly be concerned about small and marginal farmers? Looking at the agrarian history of various countries, particularly the developed countries, we find that they have undergone a rapid transformation over a period. They also started with small and marginal farmers, but today in many of these developed economies we find only large farmers and corporations dominating agriculture. They also have their small farmers but their small farmers will have a 1,000 acres or 2,000 or 5,000 acres. Basically, the farming is done there by big farmers. There is hardly any poverty in those countries. Yet, if this is the lesson of developed economies, the lesson emerging from economic transformation, why should we be excessively bothered about the future of small and marginal farmers in our country at this stage of development? This is a very important question to ask because we are raising this issue not purely on any ideological grounds in the sense that we are committed to small and marginal farmers for ever, but because of the social goals we have set for ourselves at this stage of development. The turning point in the developed countries came when, as a result of economic growth, there was a shift of labour force from agriculture to industry, into non-agricultural sector and there was a big rise in employment opportunities. In such a situation small-scale agriculture became unremunerative or unprofitable.

Agriculture is a labour-intensive activity and when labour shifts away from rural areas and wage rate rises then mechanisation is resorted to, and under mechanisation with high wage rate, only large-scale farming becomes viable and attractive. This is universally experienced in the course of economic development with the rise in wages and with the shortage of labour force for agriculture and for rural areas. The size of holding has become large, and small and marginal farmers are bound to disappear.

This is a lesson from history, but we are at a different stage of history. First of all, pressure of population on agriculture is very high. Although agriculture contributes only about 30 per cent or one-third of national product at the moment, in terms of labour force, at least double this proportion, *i.e.,* around 66 per cent is dependent on agriculture. If you take rural areas, it is much more, *i.e.,* 70 per cent to 75 per cent of total labour force is in rural areas. Our agricultural productivity is still low. There is widespread poverty. According to the latest estimates, between 35 per cent to 40 per cent of the rural population is below the poverty line. So this is the institutional and social setting in which we are talking about the prospects for small and marginal farmers. On the top of this, we have to reckon with the consequences of the on-going economic reforms, which everybody agrees, are going to be adverse for the poorer people in the short-run. All the studies done in the last four-five years show that in the immediate post-reform period the proportion of population below the poverty line has increased basically on account of the rise in prices. The rise in prices of foodgrains and essential commodities was more than 10 per cent per annum for at least four years following economic reforms which has affected poor people very adversely. The opportunities for employment generation have been limited. It is in this situation that the persistence of small and marginal farmers, their productivity, their improvement hold prospects for neutralising these adverse effects of economic reforms. People have been talking about safety nets in the wake of economic reforms in terms of employment, social insurance and some such crash programmes. But, agriculture is going to be the biggest safety net for a country like ours in the process of economic reforms by way of providing employment opportunities and by way of softening the rigour of inflation. So this is the importance of the small and marginal farm sector in the context of economic liberalisation.

Many of the East and South East Asian countries, including Peoples' Republic of China, launched economic reforms in the late seventies or early eighties. Many of these countries have at least a decade of experience in respect of economic reforms. In fact in those countries, agricultural reform in a way preceded economic reforms. They were successful in implementing land reforms. Whether it is China, Taiwan Province of China, or Republic of Korea and Japan, all these countries implemented land reforms far more thoroughly than we could do in India. Their economic reforms came later, but then even after economic reforms there has been a persistence of small and marginal farmers who got land after land reforms. There was no corporatisation of agriculture. To this date, there is no evidence of corporate farming on a big scale in these countries. Yet, these countries have shown a very high rate of growth of agriculture.

In the mid-eighties or late-eighties, in Republic of Korea, *i.e.,* South Korea, 81 per cent of holdings were small and marginal and this is a country where agricultural growth has been very high. In Indonesia, at least 60 per cent of the farms were small and marginal. It was 40 per cent in Malaysia, some 40-41 per cent in Thailand and Phillippines as against 47 per cent in India. In Bangladesh and Sri Lanka, of course, the percentage is very high, 60 per cent or more. In the fast growing South East Asian and East Asian Economies, the number of small and marginal farmers, their incidence or their proportion has been as high as in our country. Partly, it is on account of successful implementation of land reforms. For China, there is no point in giving the figure because there was communisation before economic reforms but after that land was given to small groups of peasant farmers whose holdings are more or less equal.

The per capita agricultural growth in Indonesia, Malaysia and Thailand was very high, around 2 per cent per annum during the eighties. We were not having any economic reforms in that period. Most of these economic reforms started in the nineties. We have small and marginal farmers all right, but we did not have serious economic reform, and our rate of per capita agricultural growth was negative in seventies; positive, but less than 1 per cent in the eighties. This experience shows that it is possible to achieve a very high rate of growth of agricultural output through the medium, small and marginal farmers. In fact, in Japan or even in South Korea, the implementation of land reforms immediately after the Second World War was so effective, that even to this day they are not able to consolidate the small farms into big farms or cooperatives. There is great resistance to any large scale farming in these countries, because nobody wants to give up land. In fact, Japanese agriculture and the Korean agriculture are not competitive internationally. They are giving very high subsidies to their producers. Most of them are small and marginal. They were not competitive in such a developed country as Japan, it is not possible to consolidate these into big holdings, because they are so much attached to land, because land is very scarce and land values are rising. They show that about 50 per cent of agricultural costs are accounted by machines such as tractors, harvest combines and all kinds of machinery that are used even by small and marginal farmers. Small and marginal farmers own their own harvest combines although it is not economical and they say if we have large scale farming, we can reduce substantially on this capital cost. Despite high capital cost, output growth has been high because of various subsidies. So this is another kind of situation, where following effective implementation of land reforms even at a stage of development where large farms are necessary, they are not able to corporatise agriculture. They corporatise agriculture only in small segments such as processing industries which are highly export oriented, horticulture, floriculture, animal husbandry, poultry, *etc*. but this is not a major part of agriculture.

Some states in India have already initiated proposals for removing ceiling on land holdings. The President of India has not yet given his assent to those bills, but there is a talk in a hushed up tone about the removal of ceiling on land holdings, saying that it is necessary to have large holdings in the context of economic reforms. However, the evidence particularly about the East and South East Asian countries where the dependence of labour on agriculture in rural areas has been very high, shows that it has been possible to achieve very high rates of agricultural growth, even on these small and marginal holdings. In our country the evidence is persuasive that small farmers are not in a disadvantageous position in the application of new technology. The area under small and marginal farmers has increased, their number has also increased. In fact the proportion of total area under small and marginal farms has increased and this increase has taken place in the period between 1970 to 1990, in the last 20 and 25 years. This is the period of Green Revolution, when we became self-sufficient in foodgrains and are also in a position to marginally export foodgrains in some period. All this we could realise by applying new technology and small and marginal farmers have applied these new methods vigorously and they have been active participants in the Green Revolution, which is borne out by the experience in all parts of the country, especially in the Eastern Zone and West Bengal.

West Bengal could register a rate of growth of 5 per cent per annum in rice output for nearly a decade. We could not anticipate this in the early eighties. All this has happened because of the strengthening position of small and marginal farmers through the *panchayati raj* institutions in West Bengal. In Eastern Uttar Pradesh also a very big change has occurred. A clear correlation exists between agricultural growth particularly in the areas where small and marginal farmers predominate like West Bengal and Eastern Uttar Pradesh and reduction of rural poverty. In fact, in these regions where small and marginal farmers predominate, where growth of agricultural output has been higher than the national average, the reduction of poverty, and also the rate of reduction, has been higher than the national average.

So, our own experience shows that for modernisation and for export orientation, it is not necessary to have larger sized farms. Small and marginal farmers can bring about this change. This plea for corporatisation is coming mainly because the activities they have in mind are different, not rice output or wheat output or even oilseeds but floriculture, horticulture, fisheries, shrimp farming and the related agro-processing which is, of course, highly capital-intensive. All these have a favourable export demand. Incidentally, in the present phase of development, it is reasonable to expect that many of these non-foodgrain items will have a favourable demand both domestically as well as for export because as incomes of the people increase, they will demand more and more of these commodities — dairy products, fisheries, meat, milk, vegetables so on and also in a processed form, in a finished form. The value addition will be very high. These new activities will have a future and will offer large employment opportunities with a large potential for reducing poverty.

The corporate sector may be interested in these activities. Therefore, they may enlarge the area but, the real forces behind this movement for lifting ceilings on land holdings are those who have successfully sabotaged land reforms earlier. Because of the increased demand for land in the market, the small and marginal farmers would be tempted to sell and give up their land and join the ranks of landless labour force. With no employment opportunities emerging in the non-agricultural sector and corporate sector not being in a position to offer large employment opportunities, the poverty situation is bound to worsen and the social tensions are bound to accentuate. This is not the kind of social goal we have in mind.

To cope with the adverse consequences of economic reforms and to reduce poverty, the approach document has come out with the figure of 4.5 per cent growth for agriculture in the Ninth Five Year Plan. Almost all the Five Year Plans have targeted 4 per cent per annum, but have been able to achieve much less. It is doubtful whether 4 per cent rate of growth can be achieved through the corporatisation of agriculture. Let us make a clear distinction between labour-intensive food crops and highly technology-intensive and capital-intensive agro-processing based products. The latter form a very small percentage. They have a very small weight in the overall agricultural output and bound to have a small weight for a long time to come. If China was able to achieve 6 per cent rate of growth for a decade and if all these South East Asian countries have been able to achieve 4 to 5 per cent for such a long period and if West Bengal was able to achieve 5 per cent, it is doubtful whether they would have been able to do, through the corporate sector. Among the economists it is now well recognised that small holders have tried to maximise output per unit of their land by maximising employment, whereas the capitalist farmers, the corporate enterprises, are not as much bothered about output per unit of land. They are mostly concerned with profit per unit of capital invested. So the kind of growth of output that we are talking about or that we have experienced in East and South East Asia would not have been possible under the corporate sector alone. It was possible because of peasant sector playing a dominant role in the growth of agricultural output. So, for quite some time we have to depend upon small and marginal farmers as an important instrument, for bringing about a high agricultural growth, so as to reduce poverty and to achieve high overall growth.

What then are the policy measures needed? What should be the perspective? The first thing is to strengthen the land base of small and marginal farmers. There is an active land market even today. There are people who are selling land, there are people who are buying land. However, only those who have resources can buy land and those who are hard pressed for money want to dispose of their land. There is no reason why we should not strengthen the position of small and marginal farmers in the land market. The National Commission on Rural Labour had recommended that the position of small and marginal farmers should be strengthened in the land market by extending to them liberal loans for purchase of land. When we are prepared to give one lakh rupees as loan at a concessional rate for the purchase of tractor, can't we give 10,000-15,000 rupees loan at some concessional rate for a small and marginal farmer to buy land? Lands are being sold and because these people are not in a position to buy land, it is going to those big farmers or middle farmers who are able to buy land. This can be one source for strengthening the land base of the small and marginal farmers.

It is very late in the day for anybody to argue that we should lower the ceiling further and redistribute land. We can say that, but there is no point in repeating such things in the present socio-political context. This is not practicable. Those surplus lands which have already been declared, but are pending in the court, those cases can be settled speedily in favour of the landless. Then, it is always possible to strengthen their position by allotting to them the degraded forest land. In West Bengal, in Orissa also, in China certainly, the peoples' involvement, the peasants' involvement in afforestation has been very successful. The concept of Joint Forest Management has proved to be very successful where peasants are given rights to use forest produce for 10, 15, 20 years on a long lease basis. They have the incentive to protect trees, to take care of the trees. So this will serve both the purpose of improving the standards of living of the rural poor as well as environmental protection. Environmental protection objective can thus be integrated with the poverty removal objective. So this is how we can strengthen the land base of the poor.

To make them viable in an economic sense, it should always be possible to consolidate their holdings. The experience in East Uttar Pradesh where the holdings are very small and fragmented - perhaps the most fragmented small holdings are there in the Eastern Uttar Pradesh – the experience has shown that wherever the consolidation of holdings was done, the investment in tubewells picked up. The profitability of investment in tubewells rose enormously because of consolidation of holdings. The other way of consolidating holdings for making them bigger, is to free the lease market. The situation was very different when tenancy was banned in the country. Banning, of course, did not mean doing away with tenancy. There are people who lease-out their land, there are people who lease-in land. It is underground or concealed. But investment has been a casualty. In fact, in many cases land owners, resumed their lands because of fear of losing their ownership and those lands which were in the hands of tenants, for 2-3 years, they are reluctant to make investment because they know that it is not a long lease and land will be resumed any time. So if we want to encourage investment, we should open the lease market. In fact the latest data show that a large proportion of area leased-in is held by small and marginal farmers and quite a good part is leased out, particularly in the eastern parts of the country, by the small and marginal farmers as well as by the large farmers. But, overall, the data show that after these leasing operations, the inequality in the operational holdings is less than the inequality in the ownership holdings. In the western part of the country, like Punjab, Gujarat and some other places there is a tendency for reverse tenancy, the small farmers leasing-out and the large farmers leasing-in. This shows that if for some reasons either the employment opportunities are available for small and marginal farmers outside or new technology and high wage rate necessitate a large holding size, it is possible to have large operational holdings, despite ceiling on land ownership, because a large farmer can lease-in land. Some middle farmers or large farmers and even small farmers can make their operational holdings bigger. Even corporations who want to get into floriculture and some of these cash crops can have reasonably large operational holdings by leasing-in land. They do not have to own land. So it is possible to encourage the scope for small farmers in some regions, and also the scope for corporate farmers in some other regions, so long as leasing is free and there can be vertical integration of production and processing.

It has also been suggested that the corporations can get into agro-processing and they can enter into contract with small farmers. One advantage of corporatisation of agro-processing is that the small and marginal farmers stand to benefit by way of extension of new technology. The corporation can extend technology, can extend credit and there is also a price advantage for farmers. Whenever there is a vertical integration, that is, the processing is integrated with the production, then price fluctuations are also lower because there is a guarantee of purchase. So, this integration of processing with production can go a long way in strengthening the small and marginal farmers' position.

The modern corporate sector can always be encouraged to enter into post-harvest activities. Some of the minor irrigation projects can be undertaken wherever they are profitable by the private sector. So it should be possible in the new context to work out an arrangement where the resourceful sector, whether it is private sector in general or corporate sector, can enter agriculture, but not by displacing small and marginal farmers, but by working as complementary to the existing farming sector.

In Y.V. Krishna Rao (Ed.), *Liberalisation and the Small Farmer*, All India Kisan Sabha and International Labour Organisation, New Delhi, 1997.

III

Rural Poverty Alleviation

25

Agricultural Growth and Rural Poverty

Some Lessons from Past Experience

The size of the existing stocks of foodgrains (of about 22 million tonnes) with the government is unprecedented. The foodgrains output increased by about 2.5 per cent per annum during the period between 1964-65 and 1975-76, which is only slightly higher than the rate of growth of population. If we compare foodgrains performance between the two agriculturally good years, namely, 1970-71 and 1975-76, the rate of growth comes to 2.2 per cent per annum. On this basis, the per capita availability of foodgrains from internal sources does not show any increase. The average annual import of foodgrains, mainly wheat, has been around three million tonnes in the recent period. It was in 1975 that the imports of wheat rose sharply to seven million tonnes. Even when we adjust for this additional import of four million tonnes, the stocks are sizeable and cannot be adequately explained by the overall rate of growth of foodgrains output.

The explanation is to be sought basically in the changing pattern of agricultural growth. The composition of foodgrains output has been changing in favour of superior cereals like wheat and rice at the expense of millets and pulses. This is due partly to the technological change which has favoured wheat and rice and partly to the price policy which has, in effect, ensured support prices for these two crops. The proportion of wheat in the total foodgrains output increased from 14 per cent in 1964-65 to about 24 per cent in 1975-76. The proportion of *jowar*, on the other hand, declined from 11 per cent to 8 per cent over this period. A study by Kusum Chopra and Gurushri Swamy has shown that owing to the rise in the profitability of wheat, *Rabi* pulses, particularly gram, have lost ground. *Kharif* pulses have lost ground to irrigated crops where new technology has proved profitable.[1] Similarly, there is evidence to show that in regions like Gujarat and Maharashtra, the introduction of high-yielding varieties of *Bajra* has led to decline in area under groundnut — the most important oilseed in India. As a result, the prices of pulses and oilseeds have increased significantly. However, the increase in their retail prices has not been reflected in higher prices for producers. Besides, owing to the uncertainty arising from the fluctuations in the prices of these commodities, it is likely that the rise in the expected prices of these commodities at the farm level has not been sufficient to offset the comparative advantage of some of the cereals arising from technological change as well as support prices.

It is clear that public procurement or government demand for foodgrains has been a major factor in clearing their supplies at the going prices. Otherwise, market prices of these foodgrains would have come down sharply leading to diversion of acreage and other resources to the competing crops like pulses and oilseeds. This lopsidedness in the composition of agricultural output can be corrected only by technological changes, in the long run, in respect of pulses and oilseeds. In the short run, however, a major remedy lies with price policy affecting the level as well as certainty of prices for these crops.

Apart from the deficiency of demand owing to the low rate of investment, an important factor behind the emergence of large surpluses of foodgrains with the government is the increasing concentration of foodgrains output in the developed regions and the large farms where the income level is already high and where, therefore, the income elasticity of demand for foodgrains is low. Consequently, a large proportion of increments in output is marketed, a good part of which is sold to the public procurement agencies. The data at the state level cannot throw satisfactory light on the regional concentration of output because, in certain cases — particularly in the case of large states — a regional concentration within the state may not necessarily lead to a high proportion of output for the state concerned. Unlike millets, rice and wheat are generally grown either in the assured rainfall areas or in the irrigated pockets. The proportion of output marketed in the case of rice and wheat is higher not only because of the favourable demand from the high income urban areas but also because the output of these crops is getting concentrated in regions where the productivity level is high. Thus, change in the commodity composition of output may itself explain, in part, the rise in the surpluses.

The regional concentration of output, whether between states or within the regions of a state, is better reflected in the surpluses contributed by individual states. We are concerned here with only that part of marketed surpluses which is procured by the government. Table 25.1 shows that Punjab, Haryana and Uttar Pradesh together accounted for 60 per cent of the total foodgrains procured by the government in 1975-76 as against only about 27 per cent in 1964-65. The share of a number of states like Andhra Pradesh, Madhya Pradesh, Maharashtra, Orissa, Tamil Nadu and West Bengal has declined significantly over this period. Punjab, Haryana and Uttar Pradesh which accounted for only about 13 per cent of rice output of the country in 1975-76, contributed as much as 40 per cent of total rice procured by the government. The share of several major rice growing states like Andhra Pradesh, Madhya Pradesh, Orissa and West Bengal in rice procured by the government has declined significantly between 1964-65 and 1975-76.

Table 25.1

State-wise Percentage Distribution of Public Procurement of Foodgrains

	Rice		*Wheat*		*Total Foodgrains*	
	1964-65	*1975-76*	*1964-65*	*1975-76*	*1964-65*	*1975-76*
	(1)	(2)	(3)	(4)	(5)	(6)
Andhra Pradesh	23.2	21.9	–	–	16.1	10.3
Assam	2.4	3.8	–	–	1.6	1.8
Bihar	1.2	1.0	–	1.3	0.9	1.2
Gujarat	1.1	0.4	–	0.4	0.8	1.3
Punjab and Haryana	9.8	27.5	77.3	69.0	21.7	42.8
Jammu and Kashmir	0.6	0.7	–	0.5	0.4	0.4
Karnataka	3.1	2.9	–	–	2.8	1.5
Kerala	1.2	0.5	–	–	0.8	0.2
Madhya Pradesh	15.0	4.4	14.9	3.4	13.3	4.0
Maharashtra	4.6	0.9	–	1.4	9.2	2.5
Orissa	8.4	3.6	–	–	5.9	1.7
Rajasthan	–	0.4	1.1	4.6	3.8	3.8
Tamil Nadu	13.1	14.5	–	–	9.1	6.8
Uttar Pradesh	4.7	12.4	6.7	19.5	5.6	19.0
West Bengal	11.6	4.8	–	–	8.0	2.3
Others	–	0.3	–	–	–	0.4
All-India	**100.00**	**100.00**	**100.00**	**100.00**	**100.00**	**100.00**

Sources : 1. Government of India, *Bulletin of Food Statistics.*
2. Reports of the Agricultural Prices Commission.

Past experience has shown that surpluses with the government can co-exist with, and are in fact traceable to, increase in poverty and unemployment in large parts of the country. The changing composition of foodgrains output as well as its regional concentration has meant lower consumption of foodgrains in the rural sector as the employment generation under such a pattern would be lower. A number of studies on the performance of agriculture have shown that labour co-efficient or the increase in employment as a result of a unit increase in output is much lower in the developed pockets and among the large farms. The inverse relationship between farm size and output per acre noticed under traditional labour-intensive technology in mid-fifties has now disappeared in many of these developed pockets as a result of the introduction of new technology which is capital-intensive and has, therefore, been increasingly adopted by the large farms. This higher rate of growth among large farms has been attained by the substitution of capital for labour leading to lower share of wage income in output.[2] Another important lesson emerging from this experience is that the rising surpluses with the government have been associated with a lower rate of growth of agricultural output. This is contrary to the general assumption that greater inequality leads to higher savings, higher rate of growth of output and consequently, a high rate of surpluses. In the present case, increase in surpluses is, no doubt, due to increase in inequality but this is not associated with the higher growth rate of output. It would be interesting to investigate into this implied complementarity between growth and better income distribution under conditions of large underutilised resources like labour and land.

The comfortable position in regard to foodgrains stocks with the government may give rise to the idea that a massive rural works programme should be launched to relieve rural unemployment because there are enough foodgrains to support the demand likely to be generated by such a programme. A similar situation had, in fact, arisen five years ago and several crash programmes for employment were launched which proved to be transitory and had a large inflationary impact. This raises a fundamental question about the strategy of agricultural development consistent with the removal of rural poverty and unemployment. The alternatives in this respect are dramatically brought out by the present situation: The first alternative is the one which has been actually followed, namely, to concentrate resources in the limited prosperous areas where the income elasticity of demand for foodgrains is low and, therefore, there is an assurance of substantial surpluses which can, in turn, be used for supporting a massive rural works programme. Such a programme becomes necessary because of the particular strategy of development followed. The second alternative is to widen the base of agricultural output growth, by allocating resources among the low productivity regions and small farms where the employment potential per unit increase in output is higher and the income elasticity of demand for foodgrains is also higher. This may give a high rate of overall output growth but may not necessarily yield large marketed surpluses per unit of increase in output. This strategy may, nevertheless, yield sizeable surpluses because of higher overall rate of growth.

If agricultural output did, in fact, grow at the rate of about 4 per cent per annum — the minimum envisaged in the Five-Year Plans — employment in the production of agricultural commodities would alone have grown by at least 3 per cent per annum as the labour co-efficient (i.e., percentage increase in labour input as a result of 1 per cent increase in output) is likely to be around .75 for all the techniques (e.g., irrigation, multiple cropping and HYV, etc.) taken together. Indeed, the employment generated would be more than proportionate to increase in output if additional output is realised basically through investment in irrigation in areas where labour is abundant.

The employment generated in the process of agricultural production would be in addition to the employment that would be generated in marketing and distribution of additional output, in the production and distribution of additional inputs and consumption goods and in the service sector. The growth in employment as a result of 4-5 per cent rate of growth in agriculture would be sufficient to absorb not only the growing labour force in agriculture, but would also help to clear much of the backlog of rural unemployment and underemployment. This is because, the growth of labour force in the rural sector would be significantly less than the growth of population owing, among other factors, to the migration to urban sector and to the fact that in the initial phases of development, in the lower income areas, as the income level increases, the participation rate of the unskilled labour, particularly females, declines significantly.[3]

It is not accidental that the idea of the Employment Guarantee Scheme and massive Rural Works Programme has found strong support and is being implemented with some vigour in Maharashtra. The foodgrains output has virtually remained stagnant in this state during the last decade and the increase in the non-foodgrains output such as sugarcane has been concentrated in the prosperous belt with very low employment potential. The percentage of area irrigated to total cultivated area is only about 8 per cent in Maharashtra which is about the lowest among all the states. According to the available data, however, about one-fourth of the cultivated area can be brought under irrigation and as much as 68 per cent of this potential remains unutilised. In a recent study of agricultural performance in Maharashtra, Nilkantha Rath concludes that "(1) It is quite clear that agriculture in the state cannot register progress unless irrigation is provided over much wider areas and new seeds are suitable and economical for adaptation under unirrigated conditions. The failure of the hybrids points to the direction in which research effort may help. (2) Irrigation is very expensive, and is being used in a concentrated way to benefit only one or two per cent land and as many farmers. It is worth examining if this water could not be made available to grow seasonal crops, thereby irrigating a much wider area and stabilising agricultural production, which may help push up the growth rate."[4] It is clear that the lower rate of public investment in irrigation and misallocation of scarce irrigation water are at the root of growing rural unemployment in Maharashtra.

The Rural Works Programme designed to strengthen the land base of agriculture, e.g., soil conservation and minor irrigation through consolidation of holdings as part of the Plan for achieving the targeted growth of agricultural output would be non-inflationary in character and would provide the basis for the sustained growth of output and employment as and when such productive works are completed. But the Rural Works Programme envisaged essentially as a means to provide employment because growth of output has been slow and is likely to be slow would have an inflationary potential.

The strategy of widening the base of agricultural growth, involving the underutilised resources such as labour in the low productivity regions, would have a large employment potential and this could be generated in the process of agricultural growth. This, no doubt, requires public investment but the difference between public expenditure for Rural Works Programme and the public investment in infrastructure like rural electrification, major and medium irrigation projects is that whereas the former is totally dependent on public investment to generate employment and essentially increases the effective demand for wage goods, the latter strategy complements investment resources from the individual farmers with public investment and increases the output of wage goods while generating significant demand for them. The former strategy could be all right if the basic problem is to raise the effective demand for wage goods so as to stimulate private investment in the production of such goods. However, increase in effective demand through such a strategy stimulates production only in those areas where public infrastructure is well developed and where the existing resources may already be better utilised. The latter strategy, on the other hand, by widening the infrastructure facilities in new areas stimulates private investment among sectors where the resources of land, labour, etc., are underutilised. The capital-output ratio among such sectors is generally low because of the abundance of labour. Besides, the increase in the essential consumption by the poor may also raise their efficiency, contributing to a further decline in the capital-output ratio. Harvey Leibenstein shows that in poor economies, there may exist, upto a point, a more than proportionate relationship between increase in calory intake and increase in productivity.[5] Since much of the capital formation in a labour-intensive agriculture is of non-monetary type, done through the use of labour, especially during the off-season, it is conceivable that the lower saving ratio of the small farmers in low productivity regions could be more than compensated by the improvement in the output-capital ratio and non-monetary capital formation over wide areas, so that output could grow faster under such a strategy. Further, such a growth process ensures that gains of development accrue to the poor sections.

Public investment in agriculture may broadly be classified into two types: (a) infrastructural investment in physical assets directly made by the government such as rural electrification, major and medium irrigation projects, roads, etc., which benefit all classes of farmers roughly in proportion to the area held by them; (b) public investment in the institutional infrastructure such as institutional credit, which supports private investment and, in principle, can benefit all the classes of farmers. This can even be designed and implemented to benefit essentially the poor classes. In practice, however, such benefits accrue mainly to the upper strata, at any rate, in the initial stages of development. The impact of both the types of public investments mentioned above would be to stimulate private investment and effort. The excessive emphasis on (b) to the neglect of (a) might bias development in favour of affluent segments. This may generate increase in private investment by the developed regions and large farms in response to the rise in the prices of output. This rise in prices can be expected because of neglect of (a) as it has a larger potential for increasing output and employment by drawing upon the large complementary resources, e.g., land and labour. Experience in the last decade shows that public investment of type (a) has slowed down leading to slower growth rate of overall agricultural output and, consequently, high prices for agricultural commodities. This, in turn, has provided incentives for investment by the affluent segments which has been supported by the public infrastructure investment type (b) e.g., long-term loans for investment in private wells, fertilisers, etc. For instance, the net area irrigated from government canals increased at the rate of 3.1 per cent per annum from 7.5 to 10 million hectares between 1953-54 and 1964-65. With the adoption of the new strategy, this growth had already slowed down to 2.5 per cent per annum between 1964-65 and 1970-71, as against an acceleration in private investment in well irrigation during this period. The affluent segments in Indian agriculture may not account for more than 15 to 20 per cent of cultivated area in the country, and further, since the capital-labour ratio and capital-output ratio are higher in this segment, overall rate of growth and employment generation has been lower.

Available evidence shows that when the irrigation potential is fully exploited, the inter-state disparities in development would be lower than at present. In the case of ground water potential where a large complementarity exists between public investment in infrastructure like rural electrification and private investment, as much as 50 per cent of the potential still remains to be exploited. A study by B.D. Dhawan for Uttar Pradesh shows that inter-district variations in the level of utilisation of the ground water potential in the Gangetic plains are significantly accounted by the inter-district variations in respect of the level of rural electrification. The elasticity of ground water utilisation with respect to rural electrification programme is almost unity for the entire Gangetic Plains.[6]

To sum up the arguments in this paper, it is contended that there is a strong complementary relationship between the growth of agricultural output and employment generation in Indian agriculture because of the existence of large underutilised resources such as land and labour. This may explain why, contrary to the general belief, increase in inequality has not led to a higher rate of growth in agricultural output. Public investments should be stepped up in the rural sector in projects which can be (a) directly undertaken by the government and where leakages of funds are the minimum and (b) which have large complementarity with the private resources of the farmers, namely, land and labour in the backward regions and among the poorer farmers. Such a strategy may raise the aggregate level of savings because of the exploitation of the complementarities to the maximum between public and private investment over wide areas in the country. Further, since capital-output ratio in these areas can be expected to be lower, the overall rate of growth of agricultural output may be higher than what has been achieved so far. This may generate larger employment and income for the poorer segments. Although the proportion of output marketed per unit of output may be lower, the aggregate marketed surpluses as well as savings may turn out to be higher because of higher overall growth rate.

References

1. Kusum Chopra and Gurushri Swamy, *Pulses: An Analysis of Demand and Supply in India*, Sterling Publishers, 1975.
2. C.H. Hanumantha Rao, *Technological Change and Distribution of Gains in Indian Agriculture*, Macmillan, India, 1975.
3. *Ibid*, Chapter 3.
4. Nilkantha Rath, "Performance of Agriculture in Maharashtra, 1960-72," mimeographed.
5. Harvey Leibenstein, *Economic Backwardness and Economic Growth*, John Wiley, New York, 1957, pp 64-65.
6. B.D. Dhawan, "Rural Electrification and Its Impact on Ground Water Exploitation" (mimeographed), Institute of Economic Growth, Delhi.

Economic and Political Weekly, Vol. XII, Nos. 33-34, Special Number, August 1977.

26

Strategy for Removal of Poverty

Lopsidedness of Fifth Plan Approach

For the first time in the history of Indian planning, one finds in the Approach to the Fifth Plan the concrete objective of providing minimum consumption for the bottom 30 per cent of the population. The monthly per capita consumption of the lowest 30 per cent of the population, at 1971-72 prices, was estimated for 1973-74 to be Rs. 22.90 in the rural areas and Rs. 25.17 in the urban areas. By 1978-79, the monthly per capita consumption of this lowest 30 per cent of the population is to rise to Rs. 36.64 in the rural areas and to Rs. 39.60 in the urban areas. Thus the increase, over the Fifth Plan period from the initial 1973-74 level, works out to 60 per cent for the rural poor and 50 per cent for the urban poor.

The strategy for realising this objective has to be two-pronged. On the supply side, the volume and composition of output, consistent with these consumption targets, have to be planned for. On the demand side, the plan investments have to be such that they generate the required income for the poor. The Approach document recognises the need for both, but in its exercise it concentrates on the former only. Thus the rates of growth for different sectors and the investments needed to achieve these rates of growth have been worked out in a detailed manner. But no comparable exercise or information is presented on how the required income is to be generated for the poor. An investment plan which can achieve the required volume and composition of output does not necessarily generate the desired income for the poor. What we need is an investment plan which provides the needed goods and services as well as generates the required income for the poor to enable them to buy the goods made available.

In a poor country like India, income elasticities of demand for essential goods can be expected to be quite high even for the population above the bottom 30 per cent. Consequently, even if the supply of luxury goods is restricted, the middle and the lower-middle income groups might outbid the poorest sections and appropriate to themselves a larger share of the essential goods than is intended — unless the poor command the necessary purchasing power. The Approach paper does emphasise the need for expanding employment opportunities by encouraging labour-intensive techniques of production and by giving preference to the backward regions in the allocation of investment resources. However, unless the investment plan is translated into an income plan also, especially for the poor, there would be very little chance of achieving the desired consumption objective.

Basic Framework

The basic framework of the Plan consists in ensuring terminal year consistency amongst the output levels of the different sectors. This methodology is all right for socialist economies, from where it has been borrowed and where the desired distribution of wealth and income is already ensured through the structural changes and the institutional framework. Under those conditions, the input-output approach aiming at sectoral consistency may not seriously interfere with the distributive objectives. But such an approach is not sufficient for India, where the existing distribution of wealth and income is highly unequal and where this inequality has tended to widen in the process of economic growth.

A more disaggregative approach, involving the choice of techniques within each sector becomes necessary, if removal of poverty is to be a serious objective. In other words, in a country like India, unlike in socialist countries, there has to be a shift from the material balances approach — which is after all aggregative — to one of planned distribution of benefits of growth to the poor, in terms of employment, income and social services. Sectoral consistency has to be arrived at subject to this constraint.

The Approach paper implies that, within each of the 66 sectors considered, there would be scope for such choice among techniques and projects that the desired employment and income for the poor can be achieved with the given sectoral allocations. There is no logical basis for such a belief. If this turns out to be unfeasible empirically, then there has to be a reallocation of investment resources among these 66 sectors — by giving greater weight to the more labour-intensive sectors among them, such as agriculture.

It is possible that all the data needed to work out an income generation plan are not available, and that a neat plan for income generation can not therefore be worked out. However, mere sectoral consistency is not a fruitful solution to this problem. There has to be a reasonable assurance, at the empirical level, that there is no serious conflict between the minimum consumption objective and the allocation of investment resources planned for the different sectors. In any case, the available information does permit one to estimate the magnitude of poverty — region-wise — and to work out its income and employment implications for the poor. Such a picture would be helpful in formulating programmes of development oriented towards generation of employment and income for the poor.

Magnitude of the Problem

In what follows, we shall explore how much income needs to be generated for the poor, if the minimum consumption objective is to be achieved. The main purpose of this exercise is to indicate the magnitude of the problem and, from this, to emphasise the need for an income plan as a counter-part to the plan for achieving output targets.

We shall confine our exercise to the rural poor (bottom 30 per cent) who account for nearly 80 per cent of the poor in India. Assuming that the population would increase by about 10 per cent during the Fifth Plan period, the bottom 30 per cent of the rural population works out to at least 153 million in 1978-79. Some of them may migrate to urban areas, but would nevertheless belong to the bottom 30 per cent who will have to be provided the minimum consumption. On the basis of the desired increase in the expenditure of this class, to the extent of Rs. 13.74 per capita per month or Rs. 164.88 per capita per annum, the additional annual income needed to be generated for them amounts to Rs. 2,526 crore.

This increase in income has to come from agriculture as well as from sources other than agriculture. Let us deal with agriculture first. The rural poor can be assumed to depend on agriculture — as self-employed small or marginal farmers — to the extent of getting 50 per cent of their income from agriculture. Since agricultural output is expected to increase roughly by 25 per cent during the Fifth Plan period, the output of these small farms can be assumed to increase by about the same rate, i.e., 25 per cent — if they can be provided with credit and inputs roughly in proportion to the land held by them. The remaining half of their (rural poor) income from agriculture is derived from doing agricultural labour. Agricultural output in India has been increasing recently, mainly through an increase in yield per acre. It has been found that the rise in yield per acre among the wage-based farms — particularly with the application of fertilisers and high-yielding varieties — is associated with an increase in labour productivity. The available data suggests that, for a one per cent increase in yield per acre, labour input rises between 0.5 per cent to 0.75 per cent. Assuming the latter figure, the income of agricultural labourers can be expected to increase by 18.75 per cent (25 × 0.75) by the end of the Fifth Plan. In view of a 10 per cent increase in population among the rural poor, the per capita income from agriculture of the small farmers would increase by about 15 per cent and that of the agricultural labourers by about 8.75 per cent. Assuming further, that the rural poor derive about 60 per cent of their income from agriculture, per capita income of the rural poor (small farmers and agricultural labourers put together) would increase by about 7 per cent (15 × 0.3 + 8.75 × 0.3) on this account. Since, according to the Approach paper, their per capita expenditure is to increase by 60 per cent over this period, agriculture would account for only about 12 per cent of the additional income needed; therefore, as much as 88 per cent of the desired additional income will have to come from sources other than agricultural production.

These sources are :

(a) plantations, livestock, forestry, fisheries;

(b) mining and quarrying;

(c) construction; and

(d) other services.

According to the Approach paper, the output in sector (a) is expected to increase by about 30 per cent, that in sector (b) by about 50 per cent, and that in sector (c) by about 40 per cent. The Plan assumes constant labour productivity, which means that employment can be expected to grow at about the same rate as output. Let us assume that the average output and employment of these four sectors increases by 50 per cent. It is difficult to say to what extent the rural poor would benefit from this growth. However, if their income and employment from these sources does grow by 50 per cent (or 40 per cent per capita) — i.e., their share in these sectors remains the same as before— their per capita income would increase by 16 per cent (40 × 0.40) on account of these sources. This gives the total of about 23 per cent from agricultural and non-agricultural sources. The remaining 37 per cent, or about 62 per cent of the desired increase in their income, will have to be provided by increasing their share in these four sectors.

Additional Jobs

This additional income (Rs. 1,566 crore) can be converted into the volume of employment needed. Assuming that each person employed earns an income sufficient to support himself and two dependants at the desired per capita expenditure level, the annual wage income per employed person would be Rs. 1,320—at a daily wage of Rs. 4.40—if the person is employed for 10 months in a year. On this basis, as many as 12 million additional jobs will have to be created for the rural poor, in sectors other than agriculture, over and above the proportionate rise in their share in the employment in these sectors.

This is in addition to what needs to be done to provide for the urban poor, for those above the bottom 30 per cent and for the educated unemployed. From the available data in the Approach paper, it is not possible to say whether the provision of 12 million additional jobs for the rural poor, over and above their normal share in these sectors, would be feasible.

As mentioned earlier, the objective of this small exercise is to indicate the magnitude of the problem in regard to the employment and income generation for the poor and to emphasise the need for a plan of income generation. As things are, there are several administrative and structural deficiencies in the economy which can undermine in the process of implementation even the most carefully worked out plans. The structural resistance would be even greater in the case of an objective like removal of poverty. In these circumstances the plan for removal of poverty has a somewhat better chance of overcoming such resistance, if the plan for employment and income-generation for the poor is spelt out at least as clearly as that for sectoral allocation of resources.

Economic and Political Weekly, Vol. VIII, Nos. 31-33, Special Number, August 1973.

27

Planning for the Removal of Poverty

Some Issues Concerning Approach and Strategy

Following the pioneering work done by Professors V.M. Dandekar and N. Rath on poverty in India in the early seventies, considerable amount of work has been done on the estimation of the number of people below the poverty line at the all-India level as well as at the State, regional and even district levels. This estimational work and the debate associated with such work has been very useful and stimulating. However, we have not paid enough attention to serious analysis of the factors explaining poverty in India across the regions and over a period of time. In the recent period, however, some such analyses have sprung up. Notable among them are works from Montek Ahluwalia, Dharm Narain, and Pranab Bardhan. The International Food Policy Research Institute has made a major effort by bringing together such scholarly works as a volume in memory of late Dr. Dharm Narain. Such studies which are of comparatively recent origin, need to be pursued further to get an indepth understanding at a disaggregated level not only of the factors accounting for differences in poverty but more importantly to identify the processes associated with a reduction in poverty. The available studies are by no means unanimous in their explanations of the incidence of poverty. However, three factors emerge which together provide a reasonably good explanation of the variations in poverty. Foremost among them is the institutional factor, that is, the extent of ownership of land per household, the distribution of such landholdings, the incidence of landlessness, the proportion of population belonging to scheduled castes and scheduled tribes.

The second is the growth of output per head, especially agricultural output. The regions experiencing a faster rate of output growth have shown a reduction in the proportion of people below the poverty line. This tendency is reinforced where growth has been biased more towards agriculture. Punjab and Haryana are clear illustrations of this experience. It is useful to remember, however, that the agrarian structure of these States is conducive to wider sharing of benefits when compared to many other regions. The growth in agricultural productivity in these western states has attracted migrant labour on a large scale from the poorest regions of our country in the east and contributed to a significant increase in the incomes of these landless poor. There is no doubt, therefore, that a faster rate of output growth, especially when it is biased towards *labour-intensive* and *wage-good sector like agriculture under favourable institutional framework,* will have considerable impact in raising incomes of the poor. The demand for labour generated by a fast rate of agricultural growth results in a rise in real wages so that wage rates are generally found to be positively related with agricultural productivity. However, where agricultural growth itself is slower than the rate of growth of labour force and the pattern of growth is biased against the use of labour, there may not be an appreciable impact on wage rates.

Prices of wage goods, particularly of foodgrains, is another factor affecting poverty ratio. Dharm Narain has shown that changes in poverty ratio in India in the last two decades are positively related to changes in consumer price index, while growth of output per head has a negative impact on poverty. It is clear from this that agricultural growth induced essentially by the rise in foodgrain prices cannot lead to a reduction in poverty. There has to be greater reliance on public investment in agricultural infrastructure and on cost-reducing technology, if agricultural growth is to result in the rise in the real incomes of the poor. It is now a well-accepted proposition that *growth by itself may not contribute to the reduction of poverty,* especially in an institutional framework characterised by sharp inequalities in the ownership of assets and access to resources and technology. It has, therefore, been rightly contended that the strategy for the removal of poverty has to concern more with the *institutional changes and other forms of direct attack on poverty such as the individual beneficiary-oriented programmes.* Whereas this proposition is sound, we need to ponder as to whether we have been able to exploit fully the potentialities for reducing poverty through the growth of output even as we are continuously trying to refashion our institutional framework.

If we look at the growth and poverty profile across the regions in our country, we will find on the one hand that the benefits of growth have not percolated to the lower levels because of adverse institutional framework, and, on the other, growth itself has been very slow because of adverse institutional framework. The latter proposition is true for large parts of the country, especially the eastern belt, including the vast Gangetic plains where a large part of the poverty population of the country lives. The basic fact, therefore, is that growth itself has been slow. The institutional and the infrastructural factors assume significance for stimulating growth itself, and, of course, simultaneously for securing benefits for the poor.

It is necessary to emphasise this aspect of growth and pattern of growth through the required changes in the institutions and public investments in the basic infrastructure because in a developing economy like ours, *growth through modernisation has to be a major source for reduction of poverty.* Here, the pattern of investment and pattern of growth or content of growth are extremely important. To put it plainly, the objective of removal of poverty needs to be sufficiently integrated with the production or growth strategy in our Plans. The existing planning methodology does take into account the demand for various wage goods corresponding to the magnitude of reduction of poverty contemplated. But our Plan modelling has to become more sophisticated to incorporate choice of techniques at the disaggregated levels, locational choices and patterns of regional development, etc., so that our exercises yield solutions regarding the sectoral investments as well as the income generation for various groups or classes of people. This is a difficult and complicated task with an enormous demand on data base as well as modelling skills. Even if this work becomes feasible, the story does not end. The whole exercise needs to be translated not only into concrete programmes but also into a policy framework to influence the production processes.

Looking at the regional profile of growth and incidence of poverty, we can group the country into four broad categories. *The four western States of the country, namely, Punjab, Haryana, Gujarat and Maharashtra, present the picture of high growth and low poverty,* with a difference that the incidence of poverty is even lower in Punjab and Haryana because of high agricultural growth in a perhaps more egalitarian institutional framework. The other extreme is represented by States in the Central and Eastern plains where growth has been slow and the incidence of poverty high. The *correlation between slow growth and high poverty should be noted.* The four southern States occupy a middle position. The hill States in the north and north-east are a category by itself, where the definition of poverty applicable to the plains may not be quite relevant. There, the institutional framework is vastly different from that obtaining in any of the other three categories. The problem in this Himalayan region is essentially of infrastructural development and integration with the rest of the economy.

To focus on the States in the eastern Gangetic Plains, the gap between the potential agricultural output even with the known technology, on the one hand, and the actual level of output, on the other, is perhaps the highest in the country. This potential can be tapped mainly through the development of infrastructure, both physical and institutional, requiring large investments as well as refashioning of institutions. A major developmental thrust by way of irrigation development, drainage, extension of new technology is called for. Simultaneously, land reform in certain regions, involving security of tenure for tenants, as well as redistribution of land, are necessary. Consolidation of existing holdings, which are highly fragmented, seems indispensable to stimulate investment in the indivisible equipment like pumpsets. This consolidation of holdings needs to be carried at two levels. One is to bring the fragmented pieces of each individual holding together. So far, the consolidation of holdings has been concerned with this aspect only. But in the context of modernisation of agriculture and the objective of strengthening the small farm economy, it would be necessary to consolidate holdings in such a way that as many small and marginal farmers as possible are brought together into viable blocks of land so that public investment in the exploitation of ground water can be channelled to their advantage and their group power is strengthened with a view to improving their knowhow, capabilities, and bargaining power.

There is not only a need but a considerable potential for integrating the objective of removal of poverty, with growth. Agriculture offers the greatest scope for such integration. In our disenchantment with growth experience of fifties and sixties, there is a tendency to *focus almost entirely on the beneficiary-oriented programmes for the removal of poverty. Such an effort cannot be successful if agricultural growth is slow and the benefits of this growth are cornered by a few.* We therefore, propose to launch a massive programme for the development of small-farm sector throughout the country in the Seventh Five Year Plan period through a comprehensive scheme of consolidation, development of minor irrigation, and by improving the group-effort of small and marginal farmers.

Given the magnitude of landlessness and poverty, there is no justification for the belief that agricultural growth by itself, whatever may be its content and pattern, can help to solve the poverty problem. We need, therefore, to simultaneously go ahead with the beneficiary-oriented programmes like Integrated Rural Development Programme, National Rural Employment Programme, and the more recently introduced Rural Landless Employment Guarantee Programme. This component of our strategy for removal of poverty is not devoid of a theoretical basis. The income elasticity of demand for products like milk, poultry and for a number of other rural crafts, is quite high. With a proper organisation of markets, they offer enormous potential for employment for the poor in the future. Secondly, these activities are not land-intensive and, therefore, typically suit the marginal holdings and the landless. Even though, the capital-output ratios for some of these activities are higher than that for crop production, they are still very much lower than for many of the small-scale industrial enterprises. In many cases, where traditional skills are available and remain underutilised for want of resources, the capital-output ratio turns out to be much lower than even for crop production. These activities in general, require government support for resources, training in skills, and marketing. The individual beneficiary-oriented programmes have come to stay but they cannot succeed in the absence of a reasonable rate of growth of agriculture and economic growth in general because the demand for these products is highly income elastic. These programmes are, therefore, highly complementary to growth in general. These programmes themselves are expected to contribute to growth in output of goods and services. Otherwise, they become inflationary, apart from becoming a source of gains for middlemen.

These programmes need to be expanded on a much larger scale during the Seventh Plan period and the banks have to play much greater role not only in providing credit but also in indentifying productive activities, providing extension services and conducting evaluations, at least on a pilot basis, so that the experience available from such projects is capable of being replicated by governmental machinery. The nationalised and commercial banks are not mere moneylenders. They need to be active agents in the fulfilment of national objective of removal of poverty in view of their resources, skills, and organisational capabilities. They should experiment with such models for the benefit of the poor.

Mere expansion of these programmes is not enough. The chain of intermediaries between the Government and the actual beneficiaries is perhaps the largest in the case of such programmes. The most important task, therefore, is to ensure that the benefits really reach those for whom they are intended. A number of changes in the design may be needed. Firstly, the allocation of resources for such programmes needs to bear some relationship to the incidence of poverty in different regions of the country. Secondly, in the choice of programmes as well as in their implementation local institutions have to be involved on an increasing scale with a view to ensuring that the activities chosen are economically viable and are in keeping with factor endowments and the resource-potential of the region concerned. Such an involvement will also help to minimise the leakages. Thirdly, to the extent possible, these activities should be organised on a group or cooperative basis so that the economies of scale inherent in some of these activities are fully realised while, at the same time, group initiative and effort of the poor is promoted. Finally, a major training programme has to be mounted to improve the skills and capabilities of potential beneficiaries.

We now come to some general issues of strategy for the removal of poverty. A major point to note, in this connection, is that the poor are likely to be engaged increasingly in activities the demand for which is highly income elastic and, therefore, one need not be pessimistic about the prospects for income generation for the poor so long as the overall economic growth is satisfactory. However, since all these activities are also capital-intensive and skill-intensive, those who have access to resources and technical services may be quick to invest in such ventures. Even now, the Government-sponsored programmes for the poor, such as dairying and poultry satisfy only a small proportion of the total market for such products, may be sometimes even less than one per cent. The remaining part of the market is still served by prosperous farmers with investible resources or other entrepreneurs whose effort would be to minimise labour-use and maximise profits. It is, therefore, important that by devising appropriate institutions for public investments as well as through the provision of various technical services on a preferential basis for the poverty sector, an increasing proportion of market for such products is captured by these programmes. The very fact that these activities are less land-using, makes it favourable for the poor. But the fact that they are capital-intensive and skill-intensive and possess economies of scale, may favour commercial enterprises of a relatively large scale, and thereby defeat the very objective of making this an instrument for the removal of poverty.

Education and training in skills for the poor has to be a major plank of the anti-poverty strategy. We all know that human capital is becoming a major source of economic growth. The way the benefits are distributed from this growth is determined by the distribution of human capabilities, i.e., education and skills. Although, investment in education and skills is determined by one's wealth position, public intervention in favour of the poor can make a big difference in this regard.

The focus of attention in the Seventh Plan period has to be on the provision of education for the poor and training in various useful skills.

In the ultimate analysis, the objective of removal of poverty can be fulfilled in the measure in which the poor themselves become conscious, improve their education and capabilities, become organised and assert themselves. But we know that a large majority of the poor are unorganised and are engaged in activities which are informal and which do not lend themselves easily to organise into groups. Therefore, public intervention and the strategy for poverty-removal has to be such as to promote group-endeavour. This is essential to realise the economies of scale as well as to improve the bargaining power. Economists were actively interested in cooperative farming in the fifties. A large volume of literature cropped up at that stage. Cooperation as a technique was conceived to improve the strength of those who are otherwise weak and scattered. However, experience has shown that cooperative farming as well as other cooperatives for the really weak and the scattered, have not been a success. The success stories in the case of cooperatives, interestingly, relate to those who are already rich and strong individually. The cooperatives have helped them to become even stronger. We need to reverse this trend. The situation now is far more favourable for group or cooperative effort on the part of the poor and the weak, as compared to three decades ago. For one thing the general level of awareness or consciousness among the poor and even their educational level is much higher now than three decades ago when economists actively debated the subject of cooperatives for the poor. Secondly, purely in economic sense, the indivisibilities to be exploited were very few in traditional activities under traditional technologies. The indivisibility-phenomenon is becoming increasingly important in many of the modern activities and even under traditional activities like agriculture, but with modern practices. The group and cooperative endeavour can offer greater economic returns now than before, partly through better use of modern equipment and technical services but also because of better access to resources and knowledge, and the pressure that one can bring to bear on seats of power through group-endeavour. This is what we call bargaining power. Therefore, we need to experiment with various forms of group-endeavour for the poor which also incidentally reduce risks for the individuals in their economic endeavour. This has to be another major plank of the strategy for the removal of poverty in the period ahead.

All these areas offer interesting and challenging opportunities for economists and social scientists in general, for research and investigation. A decade ago, one used to hear among the corridors of the seats of power — whether political or bureaucratic — that production is the first and foremost thing, and distribution will take care of itself. The situation has radically changed in the course of the last decade. The orientation now is towards work for the betterment of the poor. Studies on land reforms and agrarian change, conducted by scholars like Professor A.M. Khusro, Professor B. Sarveswara Rao, and Professor G. Parthasarathy have inspired many young scholars in the country and helped us to understand the agrarian situation better and to formulate strategies for the removal of poverty. It is also gratifying to note that subjects of research chosen in social science research in general, and in economics, in particular, in various universities have a distinct bias towards problems of the poor. I am sure, this tradition will be strengthened further.

Presidential Address delivered at the Second Conference of the APEA, Warangal, 10-11th March, 1984.

28

Amartya Sen and the Empowerment of the Poor

For Amartya Sen, Nobel Prize in Economics could not have come at a more opportune moment. Those in India who have been pleading and working, in their own way, for an economic policy framework that is balanced and humane will have reason to feel greatly encouraged and strengthened. Sen's outstanding contributions and their recognition are bound to have far-reaching consequences towards rethinking in economics. They will certainly lead to the pursuit of policies having a positive impact on the welfare of the poor all over the world.

There would regrettably be a big set-back to our efforts towards eradication of poverty if, contrary to Sen's own pronouncements, his work is interpreted in responsible quarters as amounting to a total indictment of the ongoing process of economic liberalisation in India. At the same time, it would be a travesty of history to think that Sen's prescriptions are totally new to Indian planning, in the sense that they were not incorporated seriously into our planning exercises or were not experimented by way of actual policies.

The position is far more complicated than would appear at first sight. The crux of the matter is that we continue to flounder in our efforts to implement the anti-poverty package. This is basically explained by the deficiencies in our social structure and political economy. We need to address them squarely, now that the issue of poverty eradication is back on the agenda with a big bang, thanks to the honour conferred on Sen.

It may be recalled that the strategy of 'direct attack' on poverty, as against reliance on 'trickle-down' of benefits to the poor in the normal process of economic growth, was adopted in India towards early 1970s. Thanks to the imaginative and painstaking work of Sukhamoy Chakravarty under the stewardship of Indira Gandhi, Fifth Plan exercises explicitly incorporated the poverty eradication objective for ensuring employment and income generation, or 'entitlements' to use Sen's words, for the bottom 30 per cent of the population. Variants of this have been followed up in the subsequent plan exercises.

It was during this period that the Minimum Needs Programme was launched. This Programme, pursued since then, includes, among other things, provision of elementary education, primary health care, drinking water, basic nutrition to vulnerable groups and housing for the weaker sections. Later, for effective implementation of such programmes, on Rajiv Gandhi's initiative, decentralisation of power to the local elected bodies was statutorily provided by empowering weaker sections and women through the reservation of one-third of seats for them.

If, despite all this, we failed to make a sizeable dent on poverty, we need to do some introspection on the strategy of development we had followed and also on the socio-political milieu within which the poverty alleviation programmes were implemented. To be sure, Sen's prescriptions do not amount to a development paradigm *as such*, but constitute certain highly desirable elements in a paradigm.

On the question of development strategy, it would be instructive to look at the experience of former socialist countries. They had succeeded in ensuring certain basic entitlements and capabilities *i.e.,* employment, education and health care for the hitherto deprived sections. Yet, due to their protracted insulation from the global economy and to centralised bureaucratic planning, their economic systems collapsed on their own, again giving rise to widespread poverty and deprivation in some of these countries.

Ironically, China, where there is continued absence of democracy, which Sen had very perceptively held responsible for widespread famine there in the sixties, is showing a rapid and orderly transition to a market oriented economy resulting in a rapid reduction of poverty, despite some retrogression in public health care in the immediate post-reform period, as noted by Sen. On the other hand, Russia, despite a very high level of material and human development achieved in the pre-reform period, is now beset with widespread poverty on account of the chaos amidst transition from the defective development paradigm followed earlier.

In the light of this overwhelming evidence on the failure of inward-looking centralised bureaucratic planning, one can justifiably argue that the 'underactivity' of government in the social sectors, rightly focused by Sen, can not be redressed without correcting the 'overactivity' of government in other sectors. The under-funding of social sectors can not be seen in isolation from the huge losses incurred by public enterprises and the very slow rise in total factor productivity in the economy on account of the particular strategy of development followed. Moreover, the importance of appropriate strategy of development is underlined by the fact that even where high level of human development could be achieved, as in Kerala, the overall growth in state domestic product and employment remained low.

Nevertheless, because of the neglect of social sectors and under-provision of physical infrastructure in the pre-reform period, economic reforms in India need to be sequenced properly to protect and expand the essential capital outlays and social expenditures. This has unfortunately been the weakest spot of the economic liberalisation process. The concern voiced by Sen, and others in India, in this respect is well taken. Rectification of this would require raising the ratio of tax revenues to GDP and reducing subsidies to the well-off sections.

But these tasks bearing on social change have proved to be the most difficult ones to accomplish. The promises in Jawaharlal Nehru's Tryst With Destiny continue to elude us on account of the political economy that evolved through the marriage of inequitable social structure with political democracy.

True, failure of land reforms in India had adverse consequences for entitlements of the poor, their human development and demographic transition. But it was naive to expect radical land reforms to succeed in India which could be implemented only in the course of the armed national liberation movements as in China and Vietnam or by the occupation forces under General Mac Arthur in Japan, South Korea and Taiwan.

Alternative avenues for the empowerment of the poor through employment oriented growth, universal elementary education, primary health care, abolition of child labour and the social and political empowerment of women and weaker sections needed serious attention. Planning under democracy was expected to become a vehicle for such social change. However, it was hijacked by the growing middle class which appropriated to itself a disproportionate share of benefits, *e.g.* subsidies, assets and jobs.

We have yet to exploit the enormous potential for social transformation inherent in a democratic polity. Economic liberalisation can release the creative enterprise of the middle class and prevent them from appropriating public resources. The state can increasingly concern itself with empowering the poor and the deprived through measures involving their participation. Sen's book, *Economic Development and Social Opportunity,* co-authored with Jean Drèze has several useful ideas on how to trigger off social change on equitable lines.

The Economic Times, 1 January, 1999.

29

Role of NGOs in Rural Development

The magnitude, the extent and the intensity of the rural development and the Minimum Needs Programme are so massive that it was felt that any organisation interested in the uplift of the masses should be encouraged to contribute its mite for the proper implementation of these programmes. As early as in October, 1982, the then Prime Minister Mrs. Indira Gandhi, in her letter to the Chief Ministers of the States emphasised the need for widening the role of voluntary agencies for the implementation of 20-point Programme. She also suggested that the State Governments should consider setting up of a consultative group of voluntary agencies under the chairmanship of either the Chief Secretary or the Development Commissioner. The Group should have as its members representatives of voluntary agencies which are already working for rural development and have actual presence in the rural scene. Periodic meetings of such groups would give valuable feedback to the Government on the actual implementation of the schemes and help in sorting out problems affecting the work of voluntary agencies.

The need having been felt to utilise the services of voluntary agencies for the implementation of Plan programmes, it was thought desirable that there should be a dialogue with voluntary agencies on the subject. The Planning Commission organised a meeting on the Role of Voluntary Agencies in the Implementation of Anti-poverty and Minimum Needs

Programme in the context of the Seventh Five Year Plan in May, 1984 in which voluntary organisations operating in different fields of activities and areas participated.

In fact the voluntary agencies had been all along encouraged to participate in the implementation of Plan programmes, particularly in the community development programmes, since the beginning of the Planning process. Thus, the Planning Commission and the voluntary agencies have had a fairly long interface, though no specific role was assigned to or thought of in earlier plans. Incidentally, in the Fifth Five Year Plan there was a scheme of promotion of voluntary schemes and social action programmes with the avowed objective of promoting voluntary action in rural development through encouragement of pilot projects of public cooperation by providing financial assistance to the States and Union territories and voluntary organisations. The scheme was continued in the Sixth Plan.

Over the years some of the voluntary agencies themselves have adopted new techniques of having direct interaction with the poverty groups and developed professional expertise for undertaking a large number of rural development programmes as distinguished from the earlier welfare or relief oriented activities.

In the meeting that took place in the Planning Commission, the voluntary agencies themselves suggested their participation and association in respect of all the anti-poverty and Minimum Needs Programmes such as identification and selection of beneficiaries for the IRDP, dissemination of information in respect of specific programmes aimed at benefiting the poor, demonstrations through pilot projects and dissemination of failure and success stories, organising the rural poor to enhance their bargaining strength both *vis-a-vis* market forces and the vested interests that oppress and exploit them.

The voluntary organisations chosen for actively participating in the Plan implementation would be those who have already shown their capabilities in implementing such schemes and have earned credence of the beneficiaries and the local administration. They are expected to be socially responsible. However, funds allocated to them would be subject to normal audit procedure. Incidentally, quite a large number of such voluntary agencies would be receiving funds through People's Action for Development India (PADI) which is an autonomous organisation sponsored by the Ministry of Rural Development. PADI has a fairly elaborate system of measures to ensure that funds allocated are properly utilised.

Moreover, a degree of healthy competition between the bureaucratic machinery and the voluntary agencies might bring about distinct improvement in the Plan performance. The magnitude of the programme is such that help and collaboration is required from all the sectors including the organisations of the beneficiaries themselves to ensure that there is no leakage and that the programmes are implemented for the benefit of those for whom these are meant.

Based on author's interview to the Chief Editor, *Yojana,* when he was Member, Planning Commission on the new strategy of involving actively voluntary agencies in the implementation of anti-poverty programmes during the Seventh Plan. (*Yojana,* November, 1984).

30

Decentralised Planning for Rural Development: An Overview of Experience and Prospects

Introduction

We are reaching a stage in the developmental process and the evolution of our democratic polity when decentralised planning for rural development has become both necessary and possible. The outlays on social development and poverty alleviation which are divisible in nature are becoming increasingly important as a proportion to total plan outlay and, potentially, most of them can be planned best at the local level. Secondly, the dominance of the rural rich in the elected local-level institutions, though very real in most places so far, is not an unchangeable phenomenon. As it is, the awareness on the part of the poor and the underprivileged is much better now than in the initial stages of our planned development. By introducing certain measures for social change, it should be possible to ensure an effective participation of the poor in such institutions. Such a participation has become necessary not only for better planning of various poverty alleviation programmes but, more importantly, for minimising leakages in their implementation so that the benefits of these programmes reach the poor.

Experience across States

The effective decentralisation in planning in India exists today at the state levels *vis-a-vis* planning at the central level. The effectiveness of decentralisation at this level did seem to affect growth and where institutional reforms have been successful, it did seem to make an impact on social justice also. The effectiveness of decentralisation in planning at the state level seems to depend very much on the size of the state. The smaller the size of the state, the greater seems to be its ability to take decisions quickly and implement the programmes effectively by promptly reaching the grassroots levels and responding to their felt needs. Similarly, decentralisation at the sub-state levels seems to be effective and percolation of benefits to the poor satisfactory wherever land reforms have been effectively implemented. The viability of the size of the state and structural changes brought about seem to have had a greater impact on growth and social justice than formal decentralisation of planning below the state level.

The experience of Punjab and Haryana illustrates how smaller states with progressive land tenure systems can grow faster with more equitable sharing of benefits of growth even without decentralised planning at the sub-state levels. The per capita income of these states is highest among all the states of the country and the growth rate in income has also been about the highest among states. The system of land tenures is progressive and the proportion of people below the poverty line is the lowest in the country. There is hardly any decentralised planning worth the name below the state levels nor is there any enthusiasm visible among the politicians and administrators for decentralised planning, *e.g.*, district planning in these states.[1] However, owing to the increase in demand for labour in agriculture and rural industries, the wage rates are high resulting in a greater reduction in the poverty ratio than among other states. At the sub-state level, the effective unit for decentralised decision-making seem to be the millions of farm and non-farm households whose initiative has been released on account of progressive land tenures, good infrastructure, profitable technology and responsive administration.

At the other extreme are big states like Uttar Pradesh, Bihar, Madhya Pradesh and Rajasthan where decentralisation of planning at the sub-state levels is highlighted as a felt need and one finds a visible concern among politicians and administrators for decentralised planning. However, it is precisely in these states that decentralised planning is least successful. It appears as if the very largeness of size which necessitates decentralisation militates against it. For one thing, a larger state is also, as a rule, a more powerful unit in terms of concentration of political as well as bureaucratic power. Devolution of planning functions and resources to a large number of districts would mean immediate parting with enormous power enjoyed by the politicians and bureaucracy at present as also foregoing the opportunities for the distribution of patronage. A bigger state is more prone to creeping political dissent and instability and the ruling elite can ill-afford to surrender the instruments for consolidation of its power. Secondly, since decentralisation in planning does not amount to giving complete autonomy to sub-state levels and indeed involves considerable amount of work at the state level by way of co-ordinating and monitoring development work through decentralisation, a switchover from the existing style of work of taking decisions directly to an indirect management through decentralisation among innumerable units is not an easy job. These states are thus too big and centralised to embark upon decentralisation.

These states account for the lowest per capita income among all states and their rate of growth in income has in general been lower than the national average. The proportion of people below the poverty line among these states is highest when compared to other states in the country. There is, therefore, a severe resource crunch in these states in relation to the essential requirements for infrastructure development and poverty alleviation. The scarcity of available resources also militates against their adequate sharing with units at the sub-state levels.

These states are also characterised by outmoded land tenure systems with the vestiges of semi-feudal relationships. The Panchayati Raj institutions are, therefore, easily dominated by the rural elite and, in quite a few places, even by those lacking commitment to the democratic norms and processes. In such a situation, one cannot expect genuine concerns at the grassroots levels for decentralised planning, much less for an equitable distribution of benefits to the poor and the underprivileged. The technical capabilities for planning are also lacking at the lower levels, although this cannot be considered a significant bottleneck in these states.

Whatever decentralised planning is practiced at the district level in these states, and indeed in most other states in the country, is done essentially by the bureaucracy by putting together the various schemes undertaken by the different departments within the parameters set at the state level. In quite a few cases, this can even be described as district planning done at the state level. Even planning at a regional level such as for hill areas in Uttar Pradesh is reduced essentially to the putting together of schemes undertaken by various departments at the state level and at best monitoring of the flow of resources into the region on account of such schemes.

Even states like Gujarat and Maharashtra which have pioneered decentralised planning at the district level in the country, find the area of freedom for the allocation of resources at the district level extremely limited. In Gujarat, for instance, where about 30 per cent of the state plan outlay is accounted by district-level schemes, a bulk of it is accounted by the departmental schemes and only about 20 per cent of it or about 6 per cent of the total state plan outlay is amenable to free allocation at the district level. Besides, as much as 80 to 90 per cent of the outlays on district-level schemes are accounted by the expenditure for ongoing schemes and very little is left for the new schemes to be undertaken.

A more serious limitation to the effectiveness of district planning even in such advanced states, not adequately appreciated so far, arises from the failure to reduce disparities in development between different regions. Marathwada and Vidarbha in Maharashtra, Saurashtra and Kutch in Gujarat, Telengana and Rayalaseema in Andhra Pradesh and Uttarkhand in Uttar Pradesh are some clear examples of the less-developed regions. When some of these areas were merged to constitute the linguistic states at the time of the states' reorganisation, as in the case of Maharashtra and Andhra Pradesh despite the recommendation to the contrary by the States' Reorganisation Commission, it was done with the assurance that special steps would be undertaken to bring these backward regions *at par* with the developed regions. However, the politics of planning in a democratic set-up within the state as political unit have been such that it became increasingly difficult to impose sacrifices on the developed regions to the advantage of backward regions. The evidence, on the contrary, points to the accentuation of disparities as in the case of Marathwada, Vidarbha, Rayalaseema and Uttarkhand. This has happened despite the constitutional provisions to safeguard the interests of the backward regions through the establishment of regional development boards with special powers to the governors for monitoring progress, as in the case of Maharashtra and Gujarat.

It is wrong to expect that decentralised planning at the district level by itself can go a long way in reducing regional disparities in development.[2] For one thing, the backwardness of each of these regions consisting of a number of districts, arises either from specific agro-climatic factors or from long periods of neglect before they were merged into the composite states. Because of these reasons, an integrated approach to the planning of infrastructure for each of these regions is required for removing backwardness. This is necessary even for making district planning effective by providing infrastructure which is beyond the district-level outlays. The potential district-level plan outlays account for only about 25 to 30 per cent of state plan outlay. Maharashtra shows a higher figure of around 40 per cent. This means as much as 60 to 70 per cent of outlay is planned and allocated at the state level for various infrastructural items, including particularly power and major and medium irrigation, which have a bearing on the development of individual districts. Besides, in a mixed economy like ours, the level and pattern of private investment, particularly in the non-agricultural sectors, which can be influenced by state policies, is an important determinant of development. The major infrasructural investments and measures to induce private investments can be planned better at the level of a broadly homogeneous region consisting of a number of districts.

The distribution of resources for the district-level outlays among districts in several states is broadly analogous to the Gadgil Formula for the distribution of central plan assistance to states. Thus the population of a district gets a major weightage together with some indicators of social and economic backwardness. In some cases weightage is also given to urbanisation. It is not clear, on balance, whether the backward districts end up by getting significantly larger allocations in per capita terms than the developed districts. In any case, the implied progressivity is likely to be modest considering the backlog of development in the less developed districts. Therefore, unless there is a mechanism to ensure that the flow of benefits from the remaining 60 to 70 per cent of the plan outlays has a significant degree of progression, the observed progressivity in the district-wise allocations cannot make a significant impact on the removal of backwardness. It is possible to conceive of a situation where the pattern of state-level outlays is such that it more than neutralises whatever progressivity that exists in the district-level allocations. That this is indeed so has been alleged time and again by the leaders of backward regions.

Whereas Maharashtra and Gujarat have pioneered district planning, their performance in respect of regional planning has been dismal, despite the existence of constitutional provisions in this regard. Indeed, there is evidence to believe that the move for district planning got a special impetus in Maharashtra towards the seventies when there was widespread discontent in the backward regions. District planning began to be considered seriously as an alternative for countering the 'narrow' and 'separatist' regional sentiments. The ready acceptance of the idea of district planning has its basis in the reluctance on the part of a number of states to pursue planning at the regional level. However, there is no doubt that district planning can be effective in achieving its goals only in the measure in which problems of regional backwardness are simultaneously addressed through appropriate mechanisms.[3]

It is interesting to note that states where decentralised planning at the district level is effective are precisely the states whose performance in respect of land reforms has been better. West Bengal, Karnataka and Jammu and Kashmir belong to this category. West Bengal is poised for decentralised planning down to the village level. Karnataka has embarked upon a bold legislation for delegation of planning functions to the district-level elected institutions and for creating the institutional mechanism for the devolution of resources on the lines of the Finance Commission at the national level. In Jammu and Kashmir, the whole cabinet including the Chief Minister attends the district-level meetings, conducts public hearings and sanctions the schemes on the spot. Although this cannot be a substitute for decision-making by the local-level institutions, it is certainly an improvement over the practice of taking decisions in the state capital and is an effective mechanism for expediting implementation. Because of some structural changes brought about in these states, there is reason to believe that benefits of development are likely to be shared more equitably.

Existing Position and Prospects

We have come a long way from the romanticism of the fifties in regard to decentralised planning. Planning at the regional level is conspicuous by its absence and whatever planning that is practiced at the district and lower levels is virtually divorced from the Panchayati Raj institutions, which have often been superseded despite the statutory status enjoyed by them. There are now constitutional provisions for holding elections to these bodies regularly; for devolution of resources by specifying their functions; and for adequate representation to weaker sections.

A number of assumptions underlying the current optimism need critical examination in the light of past experience. Statutory provisions are no doubt desirable but do not seem to be sufficient to effect a break from the past. We have seen that regional development boards became defunct and virtually extinct in course of time despite constitutional provisions and, in the case of Maharashtra, despite the unanimous resolution passed by the state legislature urging implementation of the constitutional provisions. Constitutional provisions cannot be a substitute for the requisite political culture and political will. If the elected state governments can be superseded on account of political exigencies, there is no basis for entertaining different expectations in regard to the local-level institutions. Further, in view of the fact that even the recommendations of a constitutional body like the Finance Commission are not mandatory on the union government, it is not reasonable to expect that statutory provisions alone would be sufficient to ensure adequate transfer of resources to these institutions.

Decentralised planning as an effective instrument for the formulation and implementation of poverty alleviation programmes has not been brought into the focus, presumably because of the belief that once the power is delegated to lower levels, implementation of these programmes will automatically improve because the elected representatives are accountable to the poor and the underprivileged who constitute the vast majority of our electorate. While this is true in the long run, a formal delegation of power is not sufficient to ensure this unless simultaneous measures are taken to bring about structural changes to weaken the socio-economic power that the rural rich wield over the poor, ensure adequate representation to the poor in these bodies through the necessary electoral reforms and improve the bargaining power of the poor themselves by organising them to activity-specific groups and associations and through the involvement of voluntary or non-governmental organisations in the formulation as well as implementation of these programmes.

Notwithstanding the dismal performance in regard to decentralised planning, the need for such planning is greater now because of the increasing importance of programmes for social development and poverty alleviation and also because of the prospects for such planning are better now on account of better availability of infrastructure at the lower levels, improvement in education levels, public awareness, *etc.* At the same time, if past experience is any guide, there should be no room for euphoria in this regard.

In addition to the statutory provisions, four major areas of action are indicated from the past experience for making decentralised planning effective. These are: (1) Suitable modifications in the centrally sponsored schemes so as to impart greater flexibility for local-level planning; (2) Measures for planning at the regional level, particularly in bigger states; (3) Structural changes to release the initiative of the rural poor by freeing them from various forms of socio-economic exploitation; and (4) Strategy to ensure peoples' participation, among other things, by involving voluntary organisations. All these measures would require considerable political will on the part of both the central and state governments. These are briefly discussed below.

Centrally sponsored schemes usually embody national concerns and, despite the opposition from certain state governments, their importance has increased in the recent period. They mainly relate to agriculture, rural development or poverty alleviation, health and education. Being divisible in nature, central assistance for such schemes together with matching outlays may account for bulk of the outlays at the district level. Assuming that such schemes are intended to meet certain national goals and are essential in the sense that these are not likely to be taken up by the state governments without the central initiative, a major reform needed for the fulfilment of the objectives of the schemes is to make the schemes only indicative from the central level, that is, to define the broad purpose of the schemes and leave the detailed planning of the schemes including target-setting and devising instruments for achieving the objectives, *etc*., to the institutions at the district level. It is extremely important for central ministries dealing with such schemes to indicate clearly the area of freedom in planning available to the local institutions. Indeed, the central ministries can ensure through these programmes that decentralised planning is practiced at the local level. On the other hand, if the central ministries insist on doing detailed planning of these schemes themselves then decentralised planning will be highly circumscribed, the objectives of the schemes may not be fulfilled and the centre's credibility in its advocacy of decentralised planning will be called into question.

Regional planning for backward areas provides the test for the credibility of state governments in regard to their avowed policy of removing regional disparities in development. As pointed out earlier, district planning by itself, cannot be effective in removing regional disparities in development without a corresponding effort at the regional level. In some cases, such as in Maharashtra, the existing constitutional provisions provide for an executive role by the Governor in the implementation of regional development programmes. Fears have been expressed that such a practice may lead to parallel centres of authority within a state which may result in friction and disruption of administration. If this is indeed likely to be the case, alternative mechanisms have to be found with the consent of the peoples' representatives of the concerned regions to undertake and monitor developmental measures at the regional level, so as to carry conviction with the people of the regions concerned. The centre, on its part, should take as much interest in regional planning as in district planning, particularly in respect of the backward and sensitive regions in bigger states.

Unlike planning at the central and state levels, planning at the district and lower level is concerned essentially with the specific schemes which benefit identifiable areas and individuals who are in direct contact with the elected representatives. What is involved at these levels is not basic policy-making or major decisions on inter-sectoral allocations but translating the plans into concrete schemes suited to local conditions, mobilisation of local resources, identification of beneficiary areas and families and minimising leakages by ensuring public vigilance and participation.

In the long run, there is no alternative to democratising the rural social structure by eliminating the exploitation of the poor and the underprivileged by the big landowners, usurious money lenders and contractors. There has been a continuous interaction between democracy, planning and socio-economic structure ever since independence. However, the experience shows that the impact of the existing socio-economic structure on the functioning of democracy and planning has been more pronounced. Planning and democracy as instruments for structural change have not been tapped adequately so far.

Towards this end, it would be necessary to pursue the effective implementation of land reforms, minimum wages and provision of institutional credit including for essential consumption. It would also be necessary to ensure remunerative prices to the small producers including the tribals for their minor forest produce by abolishing the contractor system and the middle-men. Such reforms relentlessly pursued can change the balance of socio-economic power in rural areas in favour of the vast majority of the poor and the underprivileged and thus create the necessary conditions for the effective voice for these groups in the local-level institutions.

The formal structure for the participation of the people in the form of Panchayati Raj institutions can be effective only in the measure in which the initiative and awareness of the people, particularly the beneficiaries from the local-level planning, is raised by organising them into activity-specific groups, associations and co-operatives, and by using the mass media to disseminate useful information to them. The role of voluntary or non-governmental organisations is extremely important in this respect. Although the involvement of voluntary organisations in these tasks has been accepted as a matter of policy, the progress has not been very encouraging so far due to bureaucratic inhibitions and even resistance.

The crux of the problem in raising the initiative of the poor is the lack of leadership which is pro-poor, motivated and has the courage to mobilise people against injustices. A leadership with all these qualities is necessarily scarce. Perhaps the scarcest among these leadership qualities is the ability to overcome all the procedural difficulties or formalities in dealing with the official programmes and the capacity to fight against the vested interests in the rural areas. If it is the question of only providing skills and selfless services and motivating the people for group endeavour and running their institutions, such people — teachers, engineers, doctors, scientists and technicians — are available in large numbers, thanks to the long tradition of voluntary work in the country. Government intervention can play a major role by way of simplifying rules and regulations, lending full moral and material support to individual leaders and associations against harassment from vested interests, petty-bureaucrats and police. If this support is forthcoming from the government, the leadership bottleneck will not be as formidable as it appears now.

References

1. Planning Commission, (1985), *Report of the Working Group on District Planning,* (Vol. I and II).
2. Government of Maharashtra, Planning Department, (1984) *Report of the Fact Finding Committee on Regional Imbalance in Maharashtra,* (V.M Dandekar Committee).
3. Government of India, Economic Advisory Council to the Prime Minister, (1983), *First Report on Decentralisation of Development Planning and Implementation in States,* Text published in *Economic Times,* January 14, 1984.

Economic and Political Weekly, Vol. 24, No. 8, February 25, 1989.

31

Job Reservations for Backward Classes

The implementation of the Mandal Commission recommendations for job reservations for the backward classes is a big step forward towards the democratisation of our socio-economic structure. It will facilitate upward mobility within the socio-economic ladder for the hitherto oppressed and neglected classes by improving the prospects for higher income as well as better social status. The administration and public services in general can be expected to become more sensitive and responsive to the needs of the disadvantaged sections who constitute the large majority of our population.

Not the least important, it will persuade many an upper class youth endowed with land, capital, education and status to venture into new fields by using these resources rather than bank upon the cosy and secure government jobs. In the process, they may well take the lead over the backward classes for some time in respect of professional and entrepreneurial achievements.

The three basic objections voiced against the policy of job reservations for the backward classes are: (a) by discarding merit it will perpetuate incompetency; (b) since caste rather than income level is made the basis for reservations, it would introduce new inequities by encouraging the better-off amongst the backward castes as against the genuinely poor from the backward as well as the forward castes; and (c) it will undermine the secular foundation of our society by dividing the people on caste basis.

Reservations by definition entail some preference for the disadvantaged who may otherwise lose in an open competition with those who were lucky to have had good education, training and upbringing. The relevant questions then are: how serious is going to be this sacrifice of merit as revealed by tests and interviews? What would be the likely impact of such a sacrifice of merit on the efficiency of services? And, would these inefficiencies not be neutralised by special programmes for on-the-job training for such disadvantaged sections?

Reservations in government jobs for the backward classes amount to 27 per cent when their proportion in the total population is twice as much. If educational capabilities and skills are equally distributed among all sections of population, then reservations for any section in proportion to its number in the total population is unlikely to result in any sacrifice of merit. Indeed, reservations are not needed in such a situation. Since in fact educational capabilities among the backward classes are lower than the average, it can be argued that job reservations for them in proportion to their population will unduly hurt merit. However, since job reservations actually made for the backward classes constitute only half their share in the population, there is hardly any justification for the fear that the government's decision on job reservations will undermine merit.

As for efficiency, apart from skills, it depends very much on the motivation and attitudes of those in service. It is not uncommon that teachers and doctors are not in position, especially in the far-flung rural areas, because they prefer to live in the relatively comfortable urban environment. Job reservations for the backward classes insofar as they improve attendance can bring some services at the doorsteps of the people where there are none. In view of their social background, the personnel belonging to the backward classes can be expected to be more alive and responsive to the felt-needs of the common people.

The importance of on-the-job training in raising the standard of services is now well recognised. In view of the rapid pace of the social changes, the nature of services as well as the techniques of delivery undergo continual change. As such, on-the-job training is becoming increasingly important relative to the initial learning in many occupations.

To put the 'caste versus class' criterion in the proper perspective, one has to appreciate the Constitutional position as well as the ground realities where certain 'lower' castes other than the Scheduled Castes and Scheduled Tribes are identified in the common mind as poorer and oppressed classes. As per the Constitution, job reservations can be made for the *socially and educationally backward classes* and not on the basis of the *income level*. Backwardness should be social *as well as* educational and for any class so identified, there cannot be further division according to the level of backwardness or income. In the light of this legal position brought out from several court judgements over the decades in different parts of the country where reservations have been implemented, the Mandal Commission evolved its own methodology to identify the *socially and educationally backward classes*. The indicators of backwardness evolved by the Commission were applied to each of the innumerable castes listed in the earlier censuses to identify the groups of castes or classes which can be considered socially and educationally backward.

It is clear from the methodology followed by the Mandal Commission that 'caste' is only incidental to its exercise which has been used to identify 'classes' of people who are socially and educationally backward. No one having even a remote acquaintance with our social structure can deny the strong correlation that exists between the caste hierarchy and the levels of socio-economic development. This is not to deny the need to amend the Constitution to provide for reservations for the poor among the classes which have not been considered to be socially and educationally backward. It would, however, be a mockery to attempt further reservations on the basis of the income level within the backward classes covering hundreds of castes whose classification as backward is beyond question. Given our social structure and administrative culture, the income criterion can easily degenerate into a convenient device to deprive the genuinely backward classes of the benefits from reservations.

The argument that job reservations for the backward classes will divide the society on caste lines is least persuasive. It is also hypocritical. Most parents who espouse such ideals are dead against their children getting spouses from the so-called 'lower' castes. It would be dishonest to assert that selections to the jobs at present are made purely on the basis of 'merit'. In a large number of cases, the really meritorious and deserving candidates from the lower income groups and lower castes are denied opportunities because of 'contacts' and 'influence' that the better-off and upper caste candidates bring to bear upon the selection processes. Apart from educational capabilities, the socio-political dominance of the upper caste is a powerful factor influencing the selection processes.

Candidates for elections to Parliament and the State Legislatures are set up keeping in view the caste alignments. Even where the electoral colleges comprise of highly educated and enlightened people, the elections are fought on caste lines. Ministerial positions are allocated with due regard for representation to the different castes. Even the posts of Vice-Chancellors are distributed very often on caste lines. While the caste factor is so pervasive in our society, to raise this bogey when reservations are soughts to be made for the backward classes, only betrays the defensive or protective instincts on the part of the privileged groups.

Apart from reservations for the Scheduled Castes and Scheduled Tribes, reservations for the backward classes have been there for decades now in quite a few states, especially in the south. These have not caused inefficiency in the public services, or disharmony in society. On the other hand, the services there appear to be far more efficient and responsive to the felt-needs and aspirations of the common people than in some of the poorer and populous states in the north. There are many reasons for this difference but better representation of the backward classes in public services is certainly one of the important factors. There are no serious caste tensions or social disharmony in these States. This is because over a period of time, there has been an adjustment with reservations or distribution of opportunities. One of the encouraging results of job reservations for the backward castes is that the motivations of the landed and the privileged sections have undergone a big change. Instead of clinging to their large landholdings and real estates, they are venturing into several new occupations opened up by economic growth and urbanisation, by employing their capital, skills and resources.

The growing demand for job reservations to disadvantaged sections is attributable to the very slow growth of employment opportunities and cornering of the very few available jobs by the influential and the powerful. Our education system is also partly to blame. In the long run, increase in employment opportunities in the economy and vocationalisation of education so as to prepare the people for self-employment, will contribute significantly to easing the pressure on government jobs. Insufficient educational opportunities for the poor also lead to imbalances in the job market where, in the absence of reservations, the backward classes find themselves losing the race. Special programmes of education and training for them at various levels can redress this imbalance. Removal of disabilities associated with caste is the only lasting way to avoid social disharmony and division of society on caste lines.

The Mainstream, September 8, 1990, New Delhi.

Index

F

G

H

J

M

O

P

Q

R

S

T

W

Y

Z

Other books by C.H. Hanumantha Rao

Authored Books

1. Agricultural Production Functions, Costs and Returns in India
   Asia, 1965.

2. Taxation of Agricultural Land in Andhra Pradesh
   Asia, 1966.

3. Technological Change and Distribution of Gains in Indian Agriculture
   Macmillan, 1975.

4. Unstable Agriculture and Droughts : Implications for Policy
   (Jointly with Susanta K. Ray and K. Subbarao), Vikas, 1988.

5. Agricultural Growth, Rural Poverty and Environmental Degradation in India
   Oxford, 1994.

6. Agriculture, Food Security, Poverty and Environment: Essays on Post-Reform India
   Oxford, 2005.

7. Essays on Development Strategy, Regional Disparities and Centre State Financial Relations in India
   Academic Foundation, 2005.

Edited Books

1. Reflections on Economic Development and Social Change : Essays in Honour of Prof. V.K.R.V. Rao
   (Jointly with P.C. Joshi), Allied, 1979.

2. Dharm Narain : Studies on Indian Agriculture
   (Jointly with K.N. Raj and Amartya Sen), Oxford, 1988.

3. Economic Reforms and Poverty Alleviation in India
   (Jointly with Hans Linnemann), Sage, 1996.

4. Water Resources, Sustainable Livelihoods and Eco-System Services
   (Jointly with Kanchan Chopra and Ramprasad Sengupta), Concept, 2003.

5. Andhra Pradesh Development: Economic Reforms and Challenges Ahead
   (Jointly with S. Mahendra Dev), CESS, 2003.

6. Indian Economy and Society in the Era of Liberalisation and Globalisation : Essays in Honour of Professor A.M. Khusro
   (Jointly with B.B. Bhattacharya and N.S. Siddharthan), Academic Foundation, 2005.